A History Of The First Battalion, The Somerset Ligh[illegible] Prince Albert's: July 1, 1916, [illegible] /ar

V. H. B. Majendie

Printing Statement:

Due to the very old age and scarcity of this book, many of the pages may be hard to read due to the blurring of the original text, possible missing pages, missing text, dark backgrounds and other issues beyond our control.

Because this is such an important and rare work, we believe it is best to reproduce this book regardless of its original condition.

Thank you for your understanding.

A HISTORY

OF THE

1st BATTALION

The

Somerset Light Infantry

(PRINCE ALBERT'S),

July 1st, 1916, to the end of the War,

BY

Major V. H. B. MAJENDIE, D.S.O.

(Somerset Light Infantry).

Taunton:

GOODMAN AND SON, THE PHŒNIX PRESS, NORTH STREET.

1921.

INTRODUCTION.

THE story contained in these pages gives the reader a detailed account of the part the 1st Battalion took in the Great War between the opening of the Somme offensive, 1st July, 1916, and the Armistice, 11th November, 1918.

It is written by Major Majendie, D.S.O., who commanded the Battalion throughout that period, and to whose leadership the Battalion owes so much. It was not originally intended for separate publication, but was written as matter for the Regimental History when brought up to date.

Unfortunately there must be a long delay before the Regimental History can be brought out in a form worthy of the Regiment.

Those officers who read Major Majendie's manuscript considered it a pity to let it be laid aside for possibly years, and he was persuaded to publish it.

I feel sure that all those interested in the Regiment will approve this course, and I know that they will find the volume of absorbing interest.

T. D'O. SNOW,

Colonel P.A. Somerset L.I.

Sept., 1920.

INSTRUCTIONS AS TO THE USE OF THE SQUARED MAPS.

1. The large rectangles on the map, lettered D, E, F, etc., are divided into squares of 1,000 yards side, which are numbered 1, 2, 3, etc. Each of these squares is subdivided into four minor squares of 500 yards side. These minor squares are considered as lettered *a*, *b*, *c*, *d*, *a* and *b* being the top squares, *a* on the left; *c* and *d* the lower squares, *c* on the left.

A point may thus be described as lying within Square B.6, M.5.*b*, etc.

2. To locate a point within a small square, consider the sides divided into tenths, and define the point by taking so many tenths from W. to E. along Southern side, and so many from S. to N. along Western side; the S.W. corner always being taken as origin, and the distance along the Southern side being always given by the first figure. Thus the point Z would be 63; *i.e.*, 6 divisions East and 3 divisions North from origin.

3. When more accurate definition is wanted (on the 1:20,000 or 1:10,000 scales) use exactly the same method, but divide sides into 100 parts and use four figures instead of two. Thus 0847 denotes 08 parts East and 47 parts North of origin (see point X). Point Y is 6503.

DIAGRAM

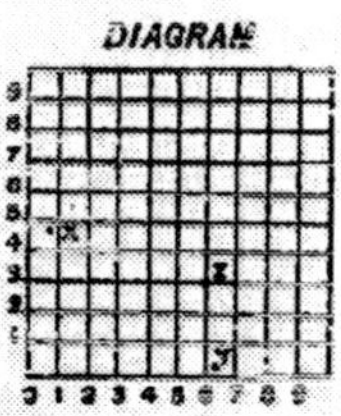

FOREWORD.

IN this attempt to present some account of the doings of the 1st Battalion, Somerset Light Infantry, in the last half of the War of 1914-1918, the intention has been to give a plain story which may be of interest to officers, W.O.'s, N.C.O.'s, and men who served with the Battalion during the War, and which may help to recall to their memory the stirring events in which they took part.

No claim is made that it is a complete and accurate historical record.

It has been written entirely from the point of view of the Battalion, and is concerned solely with the part played by the Battalion.

The 1st Somerset Light Infantry, commanded by Lieut.-Colonel E. H. Swayne, and forming part of the 11th Infantry Brigade of the 4th Division, under the command of General T. D'O. Snow, landed in France on August 22nd, 1914, and having detrained at Le Cateau at 4 p.m. on August 24th, took up an outpost line near Briastre.

The other two Brigades of the Division were the 10th and 12th. On August 26th, the 4th Division took a prominent part in the battle of Cambrai-Le Cateau, the Battalion holding a position in front of Ligny.

On the following day the retreat on Paris started, and continued until September 5th. On the 8th the allied forces advanced towards the Marne. The Battalion was engaged in the battle of the Marne on September 9th, and in the battle of the Aisne on the 13th. The 11th Brigade crossed the River Aisne on the 12th, and was the first unit of the British Army to be across.

On October 7th the 4th Division left the neighbourhood of the Aisne and entrained for St. Omer. It marched through Armentières and was engaged in the first battle of Ypres, during the course of which the Battalion fought a very successful action at Le Gheer on October 22nd.

Throughout the winter the Battalion was in Ploegsteert Wood, being almost continuously in the front line.

On December 19th an unsuccessful attack was made to improve the position, and many valuable lives were lost to the Regiment.

The Battalion did not leave Ploegsteert Wood until the middle of April, 1915, and then enjoyed eight days' rest in billets near Bailleul. It returned to the line near Pilkem, and took part in the second battle of Ypres.

On July 6th it was engaged in a very successful operation, which resulted in the capture of the Pilkem Ridge. Shortly afterwards the 4th Division left Belgium for the South, and on August 1st relieved the French in front of Beaumont Hamel.

A period of comparatively quiet trench warfare was spent in this sector until early in 1916, when the Division came out of the line for training in the neighbourhood of Beauval.

[It is regretted that some map references given in this book may appear to be inaccurate. Such discrepancies as exist are due to the fact that the maps had to be copied by hand to avoid the prohibitive cost of reproducing the original maps. Major A. J. Herington, M.C., who served for long periods with the Battalion and was second in command for the last months of the War, very kindly undertook this, and it is entirely due to his work that it has been possible to produce the maps in their present form.

Great care was taken in their production, but as was unavoidable they are not quite so accurate in every detail as the originals: this accounts for the apparent inaccuracies of some map references. It is hoped that the errors that exist are not sufficiently large to cause any real confusion, and therefore for purposes of record it has been decided in all cases to quote the real map references rather than to alter where necessary the map references to agree exactly with the maps in this book.]

CONTENTS.

LIST OF MAPS.

CHAPTER I.

THE FIRST BATTLE OF THE SOMME—THE ATTACK ON BEAUMONT HAMEL, JULY 1ST, 1916. (Map 1.)

IN the Spring of 1916 the 4th Division held the line for two months in front of Hannescamps and Fonquevillers. After three weeks' training and rest in the neighbourhood of Prouville and Yvrench, the Division moved on May 23rd to the forward area, the Battalion being accommodated in Bertrancourt.

Preparations for the forthcoming Battle of the Somme were already proceeding, and every man was required for working parties.

On June 11th the Battalion moved to Beauval for six days' training for the attack.

From now onwards everybody was working at high pressure, digging assembly trenches, carrying gas cylinders, R.E. stores, ammunition, grenades, etc., into the line. Every night the whole countryside was alive with working parties, the Serre road in particular being a mass of busy humanity. Owing to the noise and other indications, there can be little doubt but that the Germans knew what we were doing, though, except for an occasional shell or a burst of machine gun fire, they kept remarkably quiet.

On June 22nd the Battalion was ordered to move into billets in Mailly-Maillet, but the village was so congested and so heavily shelled that it was decided to bivouac in a wood just outside.

The preliminary bombardment now started, with occasional discharges of gas, but the German guns remained comparatively silent.

On June 26th C and H Companies went into the trenches under the command of Captain W. Llewellyn. The remainder of the Battalion moved into billets in Mailly-Maillet. The attack was intended to take place on June 28th, but owing to the heavy rain it was postponed until July 1st.

At 10 p.m., on June 30th, the Battalion marched out from Mailly-Maillet to its position of assembly in trenches dug in the area enclosed by Vallade Trench, Borden Avenue and Roman Road.

The ten per cent. to be left out of action with Major Majendie, Lieut. Turner, Sec.-Lieuts. De Ritter and Bennet, moved back to the reinforcement camp at Bertrancourt. On this occasion little or no provision was made for leaving out of action officers and N.C.O.'s for the purpose of reforming the Battalion in case of heavy casualties.

A few months later definite orders were very wisely issued defining exactly what officers and N.C.O.'s should be left out of a battle: included in these were the C.O. or second in command, two Company Commanders, and two Company-Sergeant-Majors. The result was that even after the heaviest casualties there was always a nucleus on which to reform.

In addition to the officers who accompanied the Battalion in the assault, Lieut. Shannon, Sec.-Lieut. Collins and R.-S.-M. Paul, M.C., were in charge of Brigade carrying parties.

General Idea of the Operations.

The attack on the VIII. Corps was carried out by the 29th, 4th and 31st Divisions in that order from South to North. The 48th Division was in Corps Reserve.

The first objective of the 4th Division was Q6c93—Q6c99—K36c35—K36a62.

The final objective allotted to the Division was R2b30—L26c35.

The 11th Infantry Brigade, under Brigadier-General Prowse, D.S.O., with the 6th and 8th Battalions of the Royal Warwickshire Regiment, of the 48th Division, attached, was given the task of capturing the 4th Division's first objective.

After this had been captured, the 10th and 12th Infantry Brigades were to pass through the 11th Brigade and capture the final objective—the 10th Brigade on the right, the 12th Brigade on the left.

The 11th Brigade attacked with three battalions in the front line—from right to left, 1st East Lancashire Regt., 1st Rifle Brigade and 8th Royal Warwickshire Regt.—to capture and consolidate the line Q5c91—Q5b17—K35d15—K35d49—K36ac5.

Three battalions formed the second line—from right to left, 1st Hampshire Regt., 1st Somerset Light Infantry, and 6th Royal Warwickshire Regt.—with orders to advance through the leading battalions and capture the first objective of the Division Q6c93—Q6c99—K36c35—K36a84.

The frontage allotted to the Somerset L.I. was bounded on the right by a line through Q5a29—K35c60—K35d60—Q6a75.40, and on the left by a line through K35a30—K35a60—K35a85.15—K35d49—K36c36.

The Attack of July 1st, 1916, at Beaumont Hamel.

July 1st was fine and warm. After an intense bombardment a large mine was exploded under the Hawthorn Redoubt at 7.20 a.m. Fortunately the Battalion had very few casualties while waiting in the assembly trenches, and all ranks were in the highest spirits, eagerly looking forward to zero hour.

At 7.30 a.m. the attack was launched. The 11th Brigade advanced in magnificent style, and the formations were accurately kept.

The Battalion advanced on a one company frontage in four lines, each line in lines of sections. A and B were the leading companies, with C and H in support. The advance started excellently, and the enemy barrage was not severe.

Very soon, however, heavy rifle and machine gun fire was opened from both flanks. The 1st East Lancashires and the 1st Hampshires were unable to get beyond the enemy wire.

The Battalion was forced to swing to its left, owing to the rise in the ground, which it should have crossed, being swept by machine guns and quite impassable, and found itself in the German trenches in the neighbourhood of the Quadrilateral.

The Warwicks on the left appear to have gained their objective, but were unable to hold on there.

The 4th Division was greatly handicapped by the 31st Division being unable to make any progress in front of Serre and by the 29th Division on the right being unable to capture Beaumont Hamel.

It is impossible to give a detailed account of the fighting that ensued, but the situation after the first two hours was that men of various battalions in the Division were holding part of the Quadrilateral, and were engaged in a fierce grenade fight. Elsewhere our men were back in the trenches from which they had started. Owing to the very heavy casualties and the small number of survivors from this action, it was extremely difficult to discover exactly what happened in the first few hours of the battle. Also the appearance of the ground had greatly altered owing to the prolonged and intense bombardment by our artillery; whole trenches had been blotted out, and it was a matter of the greatest difficulty for any individual to locate his exact position.

It is worthy of note that during the morning a report was received at Divisional H.Q. from an aeroplane that men of the Battalion had reached their objective, and were in Munich Trench. If this report was accurate, it is improbable that any of these ever returned.

Our men in the Quadrilateral were for some time severely handicapped by a shortage of grenades, but a supply was subsequently sent across mainly owing to the gallantry and good leadership of Regtl.-Sergt.-Major E. Paul, M.C., who was in charge of the carrying party. By this time the only officers left with the remnants of the Battalion were Captain A. J. Harington, M.C., and Lieut. G. C. V. Greetham. Both these officers left wounded about 1.30 p.m., and Coy.-Sergt.-Major Chappell was then in command of our men in the German trenches.

Quite early in the attack the six commanding officers of the Brigade became casualties, and after Brigadier-General Prowse was wounded, the Brigade-Major, Major Somerville, took command.

About 4.30 p.m. the officers who had been left out of action arrived in our original front line trenches, and took command of the survivors, who had been collected and re-organised in the assembly trenches by Regtl.-Sergt.-Major Paul.

After dark those men of the 11th Brigade, who were in and around the Quadrilateral, were relieved by the Royal Irish Fusiliers of the 10th Brigade, and withdrew to our own lines.

Later in the night it was decided to evacuate the small portion of the German trenches that remained in our possession, and by dawn on July 2nd the British line on this part of the front was exactly the same as it had been before the attack.

Orders were received about 10 p.m. for the 11th Brigade to move back into Divisional Reserve in Mailly-Maillet, and the line to be held by the 10th and 12th Brigades.

The losses of the Battalion were exceedingly heavy. With the exception of the Medical Officer, Captain Acland, R.A.M.C., and Sec.-Lieut. Marler, Brigade dump officer, no single officer, including those in charge of carrying parties, who formed up in the assembly trenches remained for duty at the end of the day. Lieut.-Colonel Thicknesse and the Adjutant, Captain Ford, were both killed before no-man's land was reached. The Battalion's casualties were 26 officers and 438 other ranks. The detail of officers' casualties is given below :—

Killed—

Lieut.-Col. J. A. Thicknesse.
Capt. & Adjt. C. C. Ford.
Capt. R. J. R. Leacroft, M.C.
Sec.-Lieut. G. P. C. Fair.
Sec.-Lieut. H. E. Whitgreave.

Died of Wounds—

Sec.-Lieut. H. L. Colville.

Missing, and of whom nothing has since been heard—

Capt. G. H. Neville, M.C.
Lieut. E. C. MacBryan.
Sec.-Lieut. J. A. Hellard.
Sec.-Lieut. J. A. Johnston.
Sec.-Lieut. A. V. C. Leche.
Sec.-Lieut. P. E. Dunn.
Sec.-Lieut. W. H. Treasure.
Sec.-Lieut. F. A. Pearse.
Lieut. V. A. Braithwaite, M.C.
Sec.-Lieut. G. C. Winstanley.
Sec.-Lieut. T. M. Doddington.

Wounded and Captured—

Sec.-Lieut. H. M. Tilley.

Wounded—

Capt. W. W. Llewellyn.
Capt. A. J. Harington, M.C.
Lieut. G. C. V. Greetham.
Lieut. C. J. O. Danbery.
Sec.-Lieut. R. C. Strachey.
Sec.-Lieut. A. H. Collins.
Sec.-Lieut. A. R. Waugh.
Lieut. R. W. Shannon.

In addition to the above, Brigadier-General Prowse, who came out to France with the Battalion and later commanded it, was mortally wounded and died shortly afterwards.

There is little more to add about this attack, which was a complete, but a glorious, failure, and in many ways as creditable to those, who took part in it, as many subsequent successes. By the light of experience gained later, there is little doubt that the lack of a creeping barrage, which at the time had not been evolved, allowed the Germans to make full use of their numerous machine guns, and accounted to a great extent for our lack of success.

The importance of systematically dealing with the German dug-outs as the advance proceeded was not at the time thoroughly realised: there were several instances of Germans emerging from their dug-outs after the British had passed, and firing into their backs.

After the attack of the 11th Brigade had broken down, the 10th and 12th Brigades were ordered not to advance, but one or two of the battalions had started, before the orders reached them, and they fought alongside the 11th Brigade in and around the Quadrilateral.

The following day, July 2nd, was devoted to replacing deficiencies of equipment and reorganising the Battalion. The ten per cent. reinforcements rejoined. In addition to those who had been left out of action, two officers returned to duty, Captain C. J. Peard, who had been Town-Major of Mailly-Maillet, and Sec.-Lieut. Tanner from the Divisional School.

Lieut.-Col. J. A. Thicknesse was buried on the evening of July 3rd, in the Military Cemetery, close to the Sucrerie, East of Mailly-Maillet. Brigadier-General H. C. Rees was appointed Brigade Commander.

The Battalion marched to huts at Bertrancourt on July 3rd, and returned to the front line on the 10th, relieving the 2nd Essex Regt., of the 12th Brigade, on the left of the Divisional front, just North of the area from which the attack of the Battalion on July 1st had been launched.

Heavy rain had fallen, and the trenches were in a bad state. Instructions were issued to adopt an aggressive attitude to keep the enemy occupied and to prevent him from moving reinforcements down to the South. This aggressiveness took the form of active patrolling, the discharge of smoke whenever the wind permitted, and constant artillery bombardments.

The Medical Officer, Captain J. D. Acland, unfortunately was killed on the 13th. At his own request he had gone out at night into no-man's land to search for a wounded man and was shot by a sniper.

On the 15th the Battalion, on relief by the 1st Rifle Brigade, moved back into Mailly-Maillet, with two companies in Ellis Square and the Sucrerie.

A welcome reinforcement of 16 Sec.-Lieutenants arrived from England, five from the 3rd and eleven from the 9th Battalions.

On the next evening the Division was relieved by the 36th Division, and the Battalion marched into billets at Bus.

The Brigade marched on the following day to Beauval; a short stay only was made here, and at 9 p.m. on the 20th the Battalion marched to Doullens, and entrained for the North. It detrained at Esquebec at 6.45 a.m. the next morning, and proceeded to billets in Wormhout.

CHAPTER II

YPRES SALIENT—LAST PHASE OF THE BATTLE OF THE SOMME—WINTER TRAINING—WINTER 1916-1917 IN THE SOMME MUD.

EXCELLENT accommodation was found in Wormhout, mostly in scattered farms, and it was with regret that the Battalion entrained for Poperinghe at 8 a.m. on July 25th. The transport moved by road.

"P" Camp, 2½ miles outside the town, was taken over from the 2nd Irish Guards.

The next day officers carried out a reconnaissance, and at night the Battalion relieved the 1st Welsh Guards in support on the Canal Bank in front of Elverdinghe, on the extreme left of the British line and in touch with the French. On the following day the front line was reconnoitred, and at night the relief of the Scots Guards was carried out.

The Guards' Division had suffered considerable casualties from shelling previous to the relief, but during our stay in this sector the enemy's fire slackened, and fortunately few casualties were incurred.

The chief trouble was the sniping, which was persistent and accurate. The trenches almost everywhere were of insufficient depth, and the parapets not bullet-proof. Those who were tall became distinctly envious of their shorter comrades, and had to practise moving in a crouching position.

In this sector there was an interesting no-man's land, intersected by numerous old trenches. Earlier in the year the British had been forced to give some ground as the result of a German attack, but owing to their bad condition the Germans did not occupy some of the abandoned lines, and it was not worth our while to re-occupy them.

On the left the trenches were quite close together, in the centre there was a considerable distance between the two lines, and on the right the situation was somewhat unusual. The right company held a portion of the original British line, which was virtually a glorified sap, as it ended completely in the air.

The rear and left flank of this Company were uncomfortably exposed.

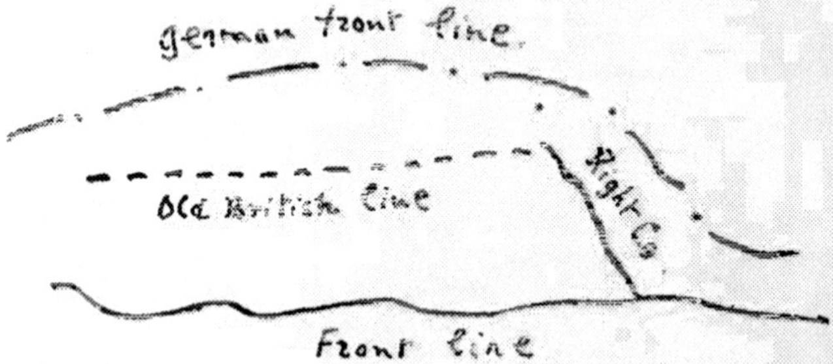

This bit of trench had to be held for purposes of observation, and to deny a rise in the ground to the enemy.

Active patrolling was carried out, and various attempts were made, but without success, to ambush German patrols, some of which were showing considerable enterprise. For some days the situation remained normal, the Battalion relieving and being relieved by the 1st Rifle Brigade, and being either in the front line or in support on the canal bank.

In the meantime the Battalion was beginning to recover from the effects of the disastrous battle of July 1st; numerous reinforcements had been received, and in addition to those officers who reported at Mailly-Maillet, several others had joined, including Major R. W. Waddy, who became second in command, Captain J. F. C. Mordaunt, Captain E. W. Marshall, and Captain D. C. H. Edwards.

But on August 8th another disaster occurred, which retarded our reorganisation and caused many casualties. On this night the Battalion was being relieved in the front line by the Rifle Brigade. In the middle of the relief, about 10.30 p.m., a gas attack was made with dense and powerful clouds of gas.

The men were all in marching order, and the noise of troops moving along the corduroy prevented the hissing of the escaping gas being heard, and this enabled the gas to come as a surprise. Incidentally the "gas alert" had been taken off earlier in the day by the Higher Command. The P.H. helmet was still in use on this date.

The gas was discharged opposite the centre company and also from saps in front of the right company at a very short range. H Coy., which was holding the right trench, had the heaviest casualties, and had in fact very few survivors. The discharge of gas was accompanied by a heavy bombardment, but the shell-fire accounted for only a few of the casualties, and no infantry followed up the attack.

The gas used appeared to be a mixture of chlorine and phosgene. This was the first occasion on which phosgene had been employed by the Germans, and nothing was known of its deadly after-effects. A few instances of these are of interest :—

Two officers of H Company, Sec.-Lieuts. Thompson and Le Peton, left the front line just before the gas attack, and on arriving at the canal bank reported at Battalion H.Q. They stated that they had had a mere whiff of the gas, but that they were perfectly fit, as indeed they appeared to be. As an infantry attack was expected, they were ordered to look for men of their Company, collect them together, and report their dispositions. This they at once proceeded to do. In two hours they were both dead.

A stretcher-bearer, after carrying down a dead man, remembered on his way back that he had not removed the personal effects from the body: he accordingly ran back to do so, a distance of perhaps 150 yards. Though apparently in excellent health, he at once collapsed.

It was proved that any extra exertion on the heart, even after slight exposure to the gas, caused a collapse, which often proved fatal. The only safe thing to do was to remain absolutely quiet.

The concentration of gas was so great that the men's clothes were impregnated with it. Several men of Battalion H.Q. were affected in this way, owing to working among casualties, removing equipment and personal effects.

Certain dug-outs on the canal bank escaped the gas cloud altogether. There were several cases of men who had come from the front line through the gas with their helmets on, and so immune from the gas, and who after sleeping in the dug-outs became casualties the next day owing to the gas in their clothes.

Towards dawn on the 9th, as casualties from the front line were constantly brought down, the scene around Battalion H.Q. of men in every stage of gas poisoning was truly pitiful, and one that those who saw it will not easily forget.

The 4th Division suffered heavily from this gas attack; the 29th Division on the right also had many casualties; and the French on the left were affected, but in a lesser degree.

The 9th August was comparatively quiet, and the enemy did not follow up his success.

During the day a large number of men went down from the after-effects of gas poisoning: Captain C. J. Peard, Sec.-Lieuts. E. V. Smith, Marler and Bennett, all of whom it was thought had escaped any ill effects, were evacuated in the evening from this cause.

The Battalion's casualties were 12 officers (6 killed) and 150 other ranks.

Detail of officer casualties :—

Killed—

Sec.-Lieut. R. P. Thompson.
Sec.-Lieut. V. F. De Ritter.
Sec.-Lieut. R. C. Roseveare.
Sec.-Lieut. D. E. Sully.
Sec.-Lieut. W. J. D. Griffiths.
Sec.-Lieut. D. A. Le Peton.

Wounded—

Lieut. R. W. Ellis.
Captain C. J. Peard.
Sec.-Lieut. I. H. Palmer.
Sec.-Lieut. E. V. Smith.
Sec.-Lieut. W. E. Marler.
Sec.-Lieut. P. H. R. Bennett.

On the night of the 10th the Battalion was relieved by the 1st Royal Irish Fusiliers of the 10th Brigade, and marched from the canal bank to Elverdinghe, whence it was conveyed in motor-busses to "J" Camp, which was reached about 5.30 a.m. on the 11th.

By this time it was realised how serious were the after-effects of phosgene poisoning: accordingly orders were issued for the transport to turn out every available vehicle to accompany the Battalion in the march from the canal bank to the embussing point. In these vehicles every man who showed any signs of exhaustion or suspicion of gas-poisoning was carried, and this probably avoided a large number of additional casualties.

The Battalion remained in "J" Camp until Aug. 21st. For the first few days very little was attempted in order to ensure that the men were entirely clear of all traces of gas. A welcome draft of 120 men arrived on the 13th, including 60 Hampshire Cyclists and 30 West Somerset Yeomanry. News was also received that the King was arriving the following day, and the Battalion was selected to find a Guard of Honour of 1 officer and 100 other ranks.

Much activity ensued to ensure a smart turn-out, but it was announced the next morning that the King's route had been altered, and that the Guard of Honour would not be required.

A further draft of 80 men was received, the majority belonging to the Devonshire Regiment.

On August 20th a reconnaissance of the front line was carried out, and on the evening of the 21st the Battalion relieved the 49th Canadian Regiment in the Hooge Sector.

The left lay on the Menin Road and the right in Sanctuary Wood. The trenches, especially on the right, were in a bad state, and in many parts water-logged. The configuration of the salient was vividly brought to notice by a large German "dud" shell, which was firmly embedded in the inside of the parapet of the trench running through Sanctuary Wood, and which obviously had been fired directly from the rear of the position held by the Battalion.

Nothing of note happened during this tour in the line, but much good work was put in on the repair of trenches. There was considerable hostile shelling, but fortunately our casualties were light.

The Battalion was relieved on August 26th by the 1st Rifle Brigade, and moved back to Montreal Camp.

On the 30th Regtl.-Sergt.-Major E. Paul, M.C., who had already done much for the Battalion, and was destined to do still more, was given a Regular Commission in the Regiment, and was at once appointed Acting-Adjutant.

The Battalion returned to the front line on the 31st, and found the trenches very wet and much damaged by recent rain. During the next few days there was considerable artillery activity in conjunction with the discharge of gas on the left by the 29th Division; also there were several alarms of gas attacks by the enemy, but none were experienced on this front.

On September 6th the enemy attempted a small raid at dawn along the Menin Road, but it was easily beaten off, and one wounded prisoner of the 362nd Regiment was left in our hands.

The Rifle Brigade took over the front line on the 7th, and the Battalion moved back, A and H Companies to the Infantry Barracks in Ypres, B and C Companies to Zillebeke Bund, and Battalion H.Q. to Ypres ramparts.

On the 8th the Battalion was relieved by the 27th Australian Infantry Battalion, and entrained at 3 a.m. on the 9th for Bollezeele, arriving there about 9 a.m.

A further move was made by road to Teteghem on September 11th.

There now took place some rather remarkable and mysterious operations, which gave rise to much conjecture and bewilderment. The affair was promptly christened, "Fogging the Bosche," and was known as such for a long time afterwards.

It appeared that the Germans were apprehensive of a British landing on the Belgian Coast behind their lines. It accordingly was decided that the 11th Brigade should carry out a demonstration from Dunkerque, which, when reported to the German authorities by their spies and agents, would cause reserves to be drawn away from the battle front on the Somme to counter this move.

To this end a large number of trawlers and small craft of various descriptions were collected together in Dunkerque Harbour.

For three days in succession the various units in the Brigade marched from their billets to Dunkerque with the greatest ostentation, as complete battalions with bands and bugles playing, proceeded to the docks and their immediate vicinity, and then towards evening returned to their billets in outlying villages as quietly as possible, marching by companies and platoons.

On each day one battalion was detailed to actually embark on the trawlers: the other units spent a pleasant time on the dunes.

To start with, a complete secrecy was maintained, and no one in a subordinate position knew at all what was happening, except that there was a vague idea that a sea voyage would be undertaken in the immediate future. After the first day the real situation was quickly grasped.

After this interlude the Battalion returned to Bollezeele on September 15th.

Information was very soon received that the time had come for a second appearance in the Battle of the Somme, and after a farewell address by the Corps Commander, General Hunter-Weston, the Battalion entrained at Esquebec at 3.30 a.m. on September 17th, arrived at Longeau at 1.30 p.m., and marched to billets in Cardonette, a few miles from Amiens. The Battalion remained here until September 25th, and much useful training was put in.

On the 25th the Battalion marched to Corbie. Here training was continued at high pressure, and every effort was made to prepare for the coming battle: the weather generally was bad. While at Corbie, Sec.-Lieut. Paul, M.C., was appointed Adjutant of the Battalion.

The Brigade moved from Corbie to Méaulte on October 7th, on the 8th to Citadel Camp, and on the 9th to bivouacs near Montauban. Heavy rain fell, and the roads were deep in mud.

The weather however improved, and a pleasant time was spent in this position. The Battalion was in the open, but temporary shelters were quickly erected.

An officers' mess was established in a large shell hole: owing to the almost entire absence of hostile aeroplane bombing camp fires were possible; a considerable amount of musical talent was forthcoming, and many pleasant musical evenings were held round the fire. Training was continued; in addition a great deal of salvage work was undertaken, and some working parties had to be supplied for digging trenches.

The task allotted to the 4th Division was to capture the

village of Le Transloy, relieving in the line near Les Boeufs the 56th Division, preparatory to the attack. This Division had been in the line for a considerable time, and was much reduced in numbers; in spite of repeated attempts, it was unable to capture the jumping-off line for the attack on Le Transloy by the 4th Division. Accordingly on October 12th the 10th and 12th Brigades carried out an attack to capture this jumping-off line, but it was unsuccessful. The Battalion "stood to" ready to move forward at an hour's notice.

On the 17th the Battalion moved up in support, relieving the 1st Rifle Brigade, who went forward into the front line. B Company was attached and in support to the Rifle Brigade. It was a pouring wet night, and everybody was soaked to the skin.

At 3.40 a.m. on the next day, October 18th, the Division attacked in conjunction with the French on the right. The Battalion was not involved in the actual attack, which did not prove successful.

The incessant rain made the trenches and the ground very difficult, and the attacking troops were further handicapped by the darkness.

Heavy machine gun are was encountered: a strong point on the left of the Rifle Brigade, formed by some old gun pits, was not carried: the right of the Rifle Brigade made some progress, but was eventually driven back, and the ultimate gain was small.

The French postponed their attack until 11.40 a.m. This also made very little progress.

A Company was sent forward to support the right company of the Rifle Brigade, and to keep touch with the French. Two platoons of B Company were also sent forward to support the right of the Rifle Brigade, and to take over and improve Frosty Trench, which had been captured and held. This trench was consolidated by B Company, and a communication trench was also dug, joining up with Warwick Avenue.

The following casualties were incurred :—

Officers Killed—

Sec.-Lieut. D. J. Blacklake.

Wounded—

Lieut. F. M. Turner.

Other Ranks—14.

During the night the Battalion took over the front line from the Rifle Brigade.

The next day, October 19th, was also very wet and cold. The trenches were in a deplorable condition, and the men suffered very much from the climatic conditions.

The French attacked and captured Sailly-Saillisel. A Company on the right co-operated and maintained touch with them. Much good work was done on the trenches, and all posts were connected up.

The next day showed some improvement in the weather, for which all were truly thankful.

The German artillery was very active all day: C Company received most attention in the early morning, and had several casualties. Battalion H.Q., in an unfinished German mine-shaft, consisting only of a few steps, and the surrounding trenches was persistently shelled from 3 till 7 p.m., but little damage was done.

During the day the Germans attacked to retake Frosty Trench, but were beaten off with severe losses.

The French made a small and successful attack on the right, and touch was maintained with them.

Captain D. C. H. Edwards was wounded and evacuated.

There was a sharp frost during the night, and the men, who were wet through, suffered very much.

The 21st was bitterly cold, and the condition of the mud went from bad to worse. Sec.-Lieut. C. V. H. Blofield was evacuated in a condition of collapse.

The Battalion, less B Company, was relieved by the 1st East Lancashire Regt. at night, and marched back to Guillemont, which was reached about 2 a.m. on the 22nd. When daylight came, the state of the Battalion was found to be deplorable. While in the line grit and a sense of duty had enabled men to keep going, but now the reaction set in.

A medical inspection was held by the Regimental M.O. in consultation with the S.M.O., with the result that 2 officers and 66 men were sent to hospital suffering from exhaustion, exposure and trench feet. A further 3 officers and 80 men were unfit for duty, and were sent back to the transport lines for rest and treatment.

On the evening of the same day, the 22nd, the remainder of the Battalion, some 300 rifles, moved forward to the line and went into Brigade Reserve. The majority were unfit for duty, and under ordinary circumstances many would have been evacuated.

A most trying period had been experienced, which had taken a heavy toll of the Battalion, though there was little or nothing to show for it.

On the next day the Battalion was due to take part in a further attack, but owing to its reduced numbers and the physical condition of the men, it was incapable of doing so, and its place in the battle line was taken by the Royal Warwickshire Regt.

The following points must be remembered in this connection :—

The Battalion moved forward into the line on the 17th in pouring rain, and every one got wet through.

An attack was imminent, and all ranks were in fighting order; no provision had been made or could have been made for a stay of several days in the trenches.

In addition, the greater part of the Battalion to reach the front line had to use a communication trench, which was waist deep in water.

On top of the incessant rain and the water-logged trenches, the nights were bitterly cold, often with a sharp frost.

A strict discipline and the innate grit and determination of the British soldier alone enabled the line to be held for so long under the existing conditions.

On October 23rd the 4th Division, together with the 8th Division on the left and the French on the right, continued the attack at 2.30 p.m. Heavy machine gun fire was again met, and only partial success was obtained. The southern gun pits were captured, and a line formed about 200 yards in front of them.

The attack on the right by the French was not successful.

At 5.30 p.m. H Company was sent up to the gun pits with grenades and ammunition: the position was very uncertain, the night was dark and great caution was required, but the stores were successfully delivered.

A patrol of 20 men, under Sec.-Lieut. Holderness, was sent out to ascertain the position on the right: most useful information was obtained, which afterwards was confirmed in every detail.

The remainder of the Battalion was used for digging and carrying parties. This was another pouring wet day.

The 24th was also a wretched day. Two officers and all the men, who had been sent to the transport lines for rest, were evacuated to hospital.

After dark the Brigade was relieved by the 19th Brigade and marched back to Trones Wood, where the remainder of the night was spent.

On the following day, in heavy rain, the Brigade moved to Mansell Camp, Carnoy.

No one in the Battalion had had any real sleep for eight nights, and the luxury of tents was much appreciated. The most pressing business of the moment was removing the mud, with which everything was plastered.

On the 30th the Battalion marched to Méricourt L'Abbé, entrained for Airaines, which was reached at 6 p.m., and marched to billets in Citerne, arriving there about 10 p.m.

Another move to Cerisy was made on November 2nd: this was the final destination, where the Battalion was to carry out training and rest.

The billets were comfortable, and a pleasant and profitable stay in this area was spent for a little over a month. The inclement weather interfered to a great extent with the training, but in spite of this much good work was carried out. Classes for N.C.O.'s were formed, and the training of Lewis gunners, signallers and bombers was taken in hand. Some drafts were received; the Battalion was completely re-organised and very soon recovered from its unpleasant experience in the mud in front of Le Transloy. Sports of every kind were much in evidence. The old custom of a "light company" competition was revived, and proved a great success. A Company was the first to gain the title of Light Company.

This was continued to the end of the war; many difficulties naturally were encountered, and some competitions were spread over a long period; but great value was obtained by the keen interest taken by all, both in the sports of every kind and also in the military events, and this led to a very real *esprit de corps* within the companies.

During this period of rest a raiding party of 2 officers and 50 other ranks was formed, and proved highly successful. The idea was to train them for raids and other enterprises of a similar nature.

The men were all volunteers, took a great pride in themselves, and considered themselves the pick of the Battalion. They were organised as a separate unit, and lived and fed apart from the companies.

They very soon proved themselves the smartest body of men in the Battalion. They were given a special training to fit them for the tasks they were likely to have to carry out. It so happened that they never had an opportunity of carrying out a raid: they remained in being for a few months, and then had to be disbanded owing to the reduced numbers of the Battalion.

Brigadier-General H. C. Rees now left for a well-earned rest in England, and was succeeded by Brigadier-General R. A. Berners.

On December 7th this period of rest ended, and the Battalion marched to Oisemont, where it entrained for Morlancourt.

On the 8th it moved to Camp 111, North of Bray, and on the 9th to Camp 16, near Bronfay.

This camp had recently been taken over from the French, was in a bad state, and surrounded by a sea of mud. During this move much rain fell.

Work was at once started to put the camp in order, and a distinct improvement was effected.

On December 13th officers reconnoitred the front line, which had recently been taken over from the French by the 4th Division.

The Battalion moved in busses on the 15th to Maurepas, and from there marched into the front line, relieving the 1st Royal Warwickshire Regt., of the 10th Brigade, just North of Rancourt.

The front line ran approximately along the Bapaume-Peronne road, facing St. Pierre-Vaast Wood, with Battalion H.Q. just North of where Rancourt Church had been.

The approach to and the condition of the trenches were exceedingly bad. When handed over by the French, they were in fair condition, but the recent heavy rain had played havoc with them, and for the most part they had ceased to exist, or were quite impassable.

The front line consisted of a series of isolated posts.

The 10th Brigade, before being relieved by the 11th, had suffered heavily from "trench feet," and had lost quite a number of men drowned or missing in the mud.

It is quite impossible to describe the winter mud of the Somme battlefield to one who has never seen it.

Fortunately it was possible to benefit by the experience of the battalions of the 10th Brigade: they had held the line in the ordinary manner in considerable strength, the only result being that the number of men who became casualties, owing to the conditions, was increased, and the general efficiency of the battalions greatly reduced.

The Germans were in an exactly similar plight, and quite incapable of offensive operations.

Accordingly it was decided to hold the line far more lightly: one company held the front line, one was in support close to Battalion H.Q., and two were kept in reserve in comparative comfort in Priez Farm. In this way it was possible to relieve the front line troops every twenty-four hours.

Owing to the lack of trenches, and the lie of the land, it was not possible to visit the front line during daylight. Consequently officers from Battalion H.Q., visiting the front line, usually went up just before dawn. These officers almost invariably on the way forward found about half-a-dozen men firmly embedded in the mud, in a state of exhaustion, and incapable of extricating themselves. These men had been lost on the previous night from relieving or carrying parties. It was often a matter of considerable difficulty to pull them out, and as a general rule their gum-boots, now a general issue, were left behind in the mud.

A case is on record of a man being extricated, but minus his trousers! This may give some idea of the nature of the mud. The men rescued in this way were taken down to the Regimental Aid Post, where they received every attention, and usually were fit for duty the next day.

It was found that several of the front line posts were only a few yards from the Germans; the latter showed signs of wishing to be friendly, but were at once discouraged.

For the first day or two the enemy was fairly quiet, but afterwards there was a distinct increase in sniping, machine gun fire and shelling, possibly owing to the aggressive policy adopted by us.

On the 18th a hard frost set in: this added to the hardships of the men, who were mostly without any cover, but at the same time somewhat improved the conditions. Good progress was made in laying corduroy tracks to the

Support Company, and this facilitated movement very much.

On the night of the 19th the Battalion was relieved by the 1st Rifle Brigade, and moved back to a camp near Maurepas. This camp came in for a good deal of shelling, and trenches had to be dug at once to gain protection. The transport lines close by were severely shelled by heavy calibre guns.

The weather was cold and wet, large parties were required to work in the forward area, and altogether these few days were not very pleasant.

On the 23rd the Battalion returned to the front line, relieving the 1st Hampshire Regt., just North of the last position, with its left just South of Sailly-Saillisel. One company held the front line, one was in support, and two in reserve.

The trenches and their approaches were found to be in a terrible condition.

Much hard work was done in making dry standings for the posts, and the laying of corduroy tracks was rapidly pushed on. The weather continued to be very bad. Christmas Day was hardly a festive occasion.

Much apprehension was felt by the Higher Command that the Germans would attempt to bring about a Christmas fraternisation. To discourage any such intentions, our artillery carried out a heavy bombardment during the day: this had the desired effect, and only drew slight retaliation.

The men suffered considerably from exposure, cold and wet, but their spirits remained excellent.

The 1st Rifle Brigade took over the front line on the 27th, and the Battalion moved back to Priez Farm, where a certain amount of shelter and accommodation were found.

On the 29th the 8th Division took over the line; the Battalion, on being relieved by the 2nd West Yorkshire Regt., marched to Maurepas, and from there was conveyed in busses to Camp 12.

During the last fortnight, hardships of the most extreme nature had been encountered, but all ranks worked together to maintain the efficiency of the Battalion, and the most elaborate precautions were taken to keep the men fit. The results were most gratifying: casualties from the weather were reduced to a minimum, and cases of trench feet were practically nil, in spite of the fact that many men had their feet affected in varying degrees as a result of the experience in front of Le Transloy.

Camp 12 was far from perfect, and much work had to be done to get it into order.

Drafts were received, consisting chiefly of North Somerset Yeomanry; the Battalion was re-organised, and every effort made to overcome the effects of the mud and exposure. Physically the men had suffered considerably, and many also were afflicted with a mental lethargy, a result of the distressing conditions. This mental state had also to be combated. Owing to the bad weather and the lack of a training ground, little field work could be carried out, but much useful instruction was given in other ways.

It was ordained that January 7th, 1917, should be observed as the Christmas Holiday, and as far as circumstances permitted quite a pleasant day was passed.

The Battalion remained in this camp until January 15th, when it moved to Camp 18, close to Suzanne.

On the following day it relieved the 1st Battalion, 77th French Regiment, just South of Bouchavesnes, with Battalion H.Q. at P.C. Brioche.

There was a long march up to the line, but the relief went off without a hitch: the arrangements of the French were good, and the hand-over generally excellent. This battalion of the 77th Regt. gave the impression of being most efficient.

This sector was particularly noticeable for the extreme length of its communication trenches.

On the night of the relief excellent work was done by the transport under Lieut. G. T. Ridge. In spite of the intense darkness, considerable shelling and most difficult ground, pack animals delivered rations and water just outside Battalion H.Q. The other battalions were not so fortunate, and had to send carrying parties a long way back.

The usual method was to push the rations, water and stores up in trucks along a light railway, but owing to the relief this railway could not be used for the first few days.

This opportunity must be taken of paying a very high tribute to the Regimental transport, which, owing to the nature of its work, does not receive in this narrative the notice which it deserves. Throughout the whole period under review, first under Captain G. Blaine, M.C., and then under Lieut. G. T. Ridge, M.C., it maintained a high standard of efficiency, and many individual acts of gallantry and good work were performed by its personnel.

On no single occasion did it fail in its allotted task. Whatever the conditions, and in spite of the weather, darkness, unknown ground, and hostile shell fire, those men of the Battalion who were in the line were able to rely on their comrades in the transport to supply them with food, water and other necessaries, and their confidence was never misplaced.

Light (A) Company, owing to some cases of measles, was isolated, and did not accompany the Battalion into the line. The trenches in this sector were comparatively good, and far better than those recently vacated North of Rancourt: the enemy were on the whole fairly quiet.

A good deal of snow fell during this period: consequently all the overland tracks were clearly defined, and the German aeroplanes became very busy with photography.

On January 19th the Brigade-Major, Captain G. A. Prideaux, M.C., was killed by a shell in the front line. His death was a great loss to the Brigade. He landed in France with the Battalion in August, 1914: he subsequently became Staff-Captain, and finally Brigade-Major of the 11th Brigade.

The Battalion was relieved by the 1st Rifle Brigade on January 20th, and moved back into support in Junction Wood.

On January 23rd, during a hard frost, it returned to the front line, relieving the 1st Royal Irish Fusiliers, immediately South of the last sector, with Battalion H.Q. in an old quarry.

On the next day Light Company, now free from infection, rejoined.

During this tour the enemy showed considerable activity with his artillery and machine guns, and also by sniping. The frost still continued.

On the 27th the Battalion on relief by the 1st Rifle Brigade, moved back into support in the neighbourhood of the quarry. Large working parties had to be found for work in the front trenches, and also on dug-out construction.

On January 30th the Battalion took over again the same portion of the front line as before. The German snipers were very active, and caused several casualties. The weather remained frosty with some snow.

The Battalion was relieved on February 1st by the 2nd Duke of Wellington's Regt., and marched out to Camp 18, arriving there at 5 a.m. on the 2nd.

At midday it moved to Camp 13, where it remained in Divisional Reserve.

Training and refitting were carried out, and a few days' rest was much appreciated.

On February 11th a move was made to Camp 17, just outside Suzanne. On the 16th the Battalion relieved the 2nd Essex Regt. in the front line of the Quarry Sector.

A thaw now set in, with the result that all the trenches, previously fairly good, fell in and became full of mud and water.

On the 19th, a wet day, the Battalion, on relief by the 2nd Duke of Wellington's Regt., went into Brigade Reserve at Junction Wood.

Here all available men were required for working parties.

The Battalion was relieved on the 22nd by the 2nd Berkshire Regt., of the 8th Division, and marched to Hem Wood, where busses were met at 8.30 p.m. to convey it to Camp 117. Owing to the recent thaw the roads were in a terrible condition, the busses constantly stuck, and camp was not reached till 6 a.m. on the 23rd.

After a few hours' rest the Battalion marched to billets in Sailly Laurette.

Training was at once started, and platoon football matches were begun.

Orders were than received to move to the 3rd Army Area.

CHAPTER III.

THE SPRING TRAINING OF 1917—THE FIRST BATTLE OF ARRAS, APRIL 9TH, 1917, AND SUBSEQUENT DAYS—SUMMER TRENCH WARFARE IN FRONT OF ARRAS.

ON March 4th the move to the North started. On that day the Battalion was billeted at Allonville, on the 5th at Talmas, on the 6th at Gazaincourt, and arrived at its destination, Noeux, at 1 p.m. on the 7th.

Training for the forthcoming attack at Arras now started in earnest. Battalion attacks were constantly practised, and open warfare was the order of the day.

Rehearsals for the attack were carried out over "flagged courses." The weather unfortunately was mostly cold and stormy, with some snow.

Particular attention was paid to sports of all kinds, and a successful boxing tournament was held.

On March the 21st the Battalion moved in busses to Diéval. Here more advanced training, consisting of Brigade exercises, was carried out.

A party of officers went forward on April 2nd to reconnoitre the trenches immediately North of Arras, and the approaches to the forward area.

On April 7th, Jellalabad Day, the move forward for the coming battle was started, and the Battalion marched to Hermaville.

It was impossible to make any preliminary arrangements, but on arrival it was found that everything necessary for the celebration of the day was ready to hand, such as suitable huts, a piano, and last but not least—an adequate supply of beer.

A concert for all ranks was held in a big hut from 5.30 till 8 p.m.

Later, from 9 till 11 p.m., the officers entertained the N.C.O.'s at a musical evening.

Considering the circumstances the day was successfully commemorated.

The fact that the march forward to the battle started on Jellalabad Day was considered to be a good omen for the success of the operations.

On April 8th the Battalion moved into tents near Maroeuil: the final preparations were made, and everybody was in the best of spirits.

GENERAL IDEA OF THE OPERATIONS. (Map 2.)

The attack was planned on a wide front, extending from the Vimy Ridge on the North for some distance to the South. Owing to the German retreat earlier in the year, the right of which pivoted on Arras, the actual front of the attack had to be shortened.

As the Germans withdrew, they laid waste the land, blew up bridges and destroyed roads, and it was found impossible to bring up in time guns and ammunition, or to make the necessary preparations for the attack as far South as originally intended.

The right of the British attack extended a few miles South of the River Scarpe.

The intention was to drive a large salient into the German position, the apex being just South of the Scarpe, East of Monchy le Preux.

The 9th Division attacked immediately North of the Scarpe, with the 34th Division on its left.

These Divisions had three objectives :—

(*a*) The Black line, the German first trench system.

(*b*) The Blue line, the railway.

(*c*) The Brown line, the Point du Jour and Laurel Trench running through H 9 central.

On these objectives being captured, the 4th Division was ordered to pass through the 9th Division and capture a portion of the 4th German system, consisting of two trenches, together with the Hyderabad Redoubt, a little further to the East, throwing a defensive flank on the left to join up with the 34th Division. It thus formed the Northern face of the salient to be driven into the German position.

The Division attacked with two Brigades in the line, the 11th on the left and the 12th on the right, with its right resting on the Scarpe: to the latter was allotted the task of capturing Fampoux.

The 10th Brigade was in reserve: it was largely detailed to carry forward stores, and its eventual role was to hold the Brown line.

The 11th Brigade attacked with two Battalions in the front line and two in support.

Front line—1st Somerset L.I. on the right, 1st Hampshire Regt. on the left.

Second line—1st Rifle Brigade on the right, 1st East Lancashire Regt. on the left.

After the German fourth system had been captured, the Rifle Brigade was to pass through the Battalion and seize Hyderabad Redoubt.

The East Lancashires were to swing to the North and dig a line of posts from the left of the Hampshires to the Point du Jour, thus forming a defensive flank.

The objective allotted to the Battalion was the fourth German system between H10d70 and H11c10, secondly to push on patrols to clear the sunken road in H11c, and to assist the Rifle Brigade to capture Hyderabad Redoubt by covering fire.

As the distance to be covered in this attack was so considerable, it was decided that each Battalion should attack with three companies only, the fourth being detailed for carrying purposes.

B Company was selected to be the carrying company. The four carrying companies were brigaded under the command of Captain S. V. Washbrough.

The Battle of April 9th, 1917.

The attack of the leading Divisions took place at 5.30 a.m. At 6.5 a.m. the Battalion marched from camp and was concentrated with the rest of the Brigade in a field just S.W. of St. Catherine, in G15a.

Heavy rain fell, but later the weather cleared.

At the assembly area breakfast was issued, and then ensued a trying period of waiting, until information regarding the attack of the other Divisions was received.

In case the attack was not successful, or only partially successful, various orders had been issued as to the part to be played by the 4th Division.

It had to be prepared to reinforce any part of the line that was held up and unable to advance.

The welcome news, however, was received that the attack had been entirely successful, and that the original plan in its entirety would be adhered to.

At 10 a.m. the Battalion left the assembly area and moved in column of route to the Blue line (the railway) in H7d.

A few stray shells fell, but without causing any casualties, and it was not necessary to deploy.

The march forward was done on a compass bearing: on arrival at the Blue line, it was found that this had been carried out correctly, and the Battalion was in its allotted position.

The Battalion was formed up in the railway cutting and behind the embankment, in touch on the left with the Hampshires and on the right with the 2nd Essex Regt.

At the appointed time the march forward was continued in artillery formation on a compass bearing. Practically no shelling was encountered, and the place of assembly on the Brown line in H9c was reached well in advance of the time-table.

At zero+9.40, i.e., at 3.10 p.m., the Battalion moved forward to the assault in artillery formation, and closed up to our barrage. It was not possible to get very close, owing to the number of shells that were falling short on account of the extreme range at which our guns were firing. The enemy put down a fair barrage, which caused some casualties.

The advance was carried out excellently, and the extension took place below the crest of the hill in H9 and 10 exactly as it had been practised.

On approaching the fourth German system some machine gun and rifle fire was opened, and it was seen that the wire had not been cut in the slightest degree by our guns.

For a moment the situation looked critical, but the superior moral of the British proved the decisive factor. Our men did not show the slightest hesitation: the majority made for gaps in the wire, where tracks ran; some started to climb over the wire, while the remainder halted and opened fire on any Germans who showed themselves in the trench. This proved too much for the shaken nerves of the Germans, and in most cases they put their hands up. Some showed fight, but these were quickly killed. A large number came out of the dug-outs and surrendered.

The men forming the garrison of the front line trench were very anxious to get away: they were pushed out through the gaps in the wire, and at once ran westwards with their hands up through the advancing Rifle Brigade. No escort was necessary, none was sent, and no man showed any desire to go back with them.

The garrison of the second trench for the most part ran away, and was fired on by Lewis guns and rifles. As soon as the objective had been captured, C Company pushed forward patrols to the sunken road; some resistance, including a machine gun, was met, but it was speedily overcome, and several prisoners, including three officers, were taken. Consolidation was immediately started: the patrols were subsequently relieved by the Rifle Brigade, which passed through the Battalion and captured Hyderabad Redoubt. The companies were reorganised, and touch was established with the battalions on both flanks.

It was now realised that the Brigade had been carried rather too far to the South, and this later was rectified. It was not discovered immediately, owing to the trenches and the features of the ground corresponding almost exactly in every detail with those that should have been found.

The German guns were quick in getting on to the captured position, and a considerable amount of shelling was experienced, mostly from the direction of Gavrelle. In the meanwhile the attack as a whole had been generally successful; on the left all objectives were gained, and Vimy Ridge carried.

The 12th Brigade, just North of the Scarpe, met with severe opposition in Fampoux, and this village was not finally captured till late in the day.

South of the river a stiff resistance was encountered around Monchy le Preux, and all the objectives were not captured on this day. The result was that the point of the wedge driven into the German position on the South of the river was not as deep as had been hoped, and a complete break-through was not achieved.

About 6.30 p.m. orders were received to hold the captured trenches with one company, and to withdraw Battalion H.Q. and two companies to Effie Trench. C Company was left in the fourth German system, while Light and H moved back. Soon after midnight Light Company was sent forward again to the sunken road in H11c.

The night was miserably cold, with rain and sleet.

The Battalion's casualties were light:—

Officers killed—

Captain H. J. Tanner.
Sec.-Lieut. A. M. Hill.

Other ranks—

50 killed and wounded.

The day was completely successful, but it was a matter of keen debate later whether the success might not have been exploited further by sending forward the 10th Brigade through the 11th and 12th.

The Rifle Brigade, after capturing Hyderabad Redoubt, was subject to a certain amount of sniping, but the number of the enemy actually in front of it was probably comparatively small.

Also later in the day considerable forces of the enemy were seen to arrive in motor-busses; these might have been caught in the act of de-bussing by the 10th Brigade, if it had been pushed through without delay.

The gap, however, through which it had to pass, was a narrow one, too narrow probably to have enabled it to have achieved any lasting success, whatever losses it might have inflicted on the enemy to start with.

During the morning of April 10th the Battalion was concentrated in Hyderabad Redoubt. About midday orders were received that cavalry would pass through the Brigade, and that the Battalion would send out patrols after them to secure the line of the Gavrelle-Chemical Works road from the Inn, through the cross roads, to I7070, which point would be the dividing line between the 11th and 12th Brigades.

Orders were accordingly issued that Light and C Companies would send out two patrols, each of a platoon with Lewis guns, to make good this line. If necessary, C Company would send out a second patrol to maintain touch with the 12th Brigade. The remainder of these two companies was to be held in readiness to move forward in support of the leading patrols. H Company was ordered to be prepared to support Light and C Companies, but was not to move without orders from Battalion H.Q.

Hyderabad Redoubt was strongly wired on all sides, and few gaps existed. This complicated the difficulties of the attacking troops.

The patrols of Light Company were ordered to leave the Redoubt by the road on the northern side, and those of C Company by the road on the southern side: as soon as they were clear of the wire, the patrols were to move to the East. These were the only gaps that existed in the wire.

During the afternoon information was received that the cavalry were not going through, and that the Battalion would act as ordered without them, advancing at 6.30 p.m. The orders given above were not changed, and artillery support was promised to cover the advance.

In view of the circumstances, permission was obtained from Brigade H.Q. not to press home the attack in the event of serious opposition being met.

At 6.30 p.m. the leading platoon of Light Company went out from the northern and the leading platoon of C Company from the southern side of the Redoubt. No artillery support was forthcoming, apparently owing to telephonic communication having broken down. Very heavy machine gun fire was at once opened by the Germans, with the result that these two platoons had exceedingly heavy casualties, and practically ceased to exist. As there was no hope of achieving success, the attack was at once stopped.

On the morning of the next day, April 11th, orders were received that the 4th Division would attack at noon to secure the line Plouvain-Greenland Hill—Inn I7a28—Hyderabad Redoubt.

Order of battle from North to South—11th, 10th and 12th Brigades.

1st objective—Roeux-Gavrelle Road, and defensive flank from the Inn to the Redoubt.

2nd objective—Plouvain-Greenland Hill, and defensive flank thence to the Inn.

The Battalion formed the extreme left of the attack, and was given the task of securing the road from the Inn to the Cross Roads in I7a, both inclusive, and of forming a defensive flank from the Redoubt to I7a28 inclusive. H and Light Companies were detailed to carry out the attack, H leading with Light in support, each company in two waves on a two-platoon front. C Company was held in reserve in the Redoubt.

The defensive flank was to be formed from right to left by II, Light and C Companies.

The only assembly area allotted to the Battalion was Hyderabad Redoubt, which, owing to its size and shape, was somewhat inadequate.

It was decided that right platoons of companies should emerge from the communication trench running out from the Eastern apex of the Redoubt, and left platoons through the gap in the wire on the Northern face. This was not an easy operation.

Shortly before zero two battalions of the 10th Brigade, which had come too far to their left, poured into the Redoubt. This area was already over-crowded; consequently a good deal of confusion took place, and the platoons of the Battalion were prevented from advancing as arranged. The leading platoons did not get clear of the wire till zero+5 minutes.

By this time the enemy, who were now in considerable strength, had opened heavy machine gun fire, and progress was impossible. The right platoon at once jumped into the trench running out of the apex of the Redoubt and worked down to within 150 yards of the advanced German line. No further ground could be made, as the trench was blown in, and the gap was swept by machine guns. This platoon maintained itself here for the remainder of the day.

The attack failed, and the Division made practically no ground. The Battalion, without a doubt, escaped heavy casualties owing to the leading platoons being prevented from moving forward till after zero.

During the night a line of posts was established East of the Redoubt, from about H12a10.35 to H12c57. Touch was gained on the left with the Hampshires and on the right with the Seaforths.

On April 12th a similar attack was attempted by the 9th Division, and this also failed.

The 1st Rifle Brigade was ordered to co-operate and form a defensive flank to the North.

The 9th Division, which advanced in daylight over the crest of the hill in the open to pass through the 4th Division, was at once observed, and the Germans put down a heavy barrage before zero, and when it was still some distance behind our front line.

The Rifle Brigade, moving into position, was also observed: before zero machine guns were brought into play from Gavrelle, and stopped its advance. At zero our artillery put down an excellent barrage, which very largely neutralised the German machine gun fire, but it was then too late, and the attack had been broken.

It appears probable that it would have been successful if the 9th Division, assisted by the 4th already on the ground, had assembled in the front line during the night and attacked at dawn.

After the failure of this attack, a large number of men of the Scottish Rifles of the 9th Division occupied the right posts of the Battalion, dug in among them, and also took shelter in the Redoubt.

For the remainder of the day Hyderabad Redoubt was subjected to heavy shelling, which caused a number of casualties. After dark the Scottish Rifles were withdrawn.

At 4 a.m. on the 14th, a company of the 5th Cameronians of the 26th Brigade arrived at the Redoubt. Though no intimation to this effect had been received, its commander stated that his battalion had been ordered to dig a line of posts connecting the Redoubt with Huddle Trench to the South. Dawn was close at hand, and this company had no tools, but every endeavour was made to enable them to carry out their task. This, however, was found to be impossible, and just before dawn it was accommodated in trenches and posts on the right of the Battalion's area.

At 5 a.m. the other three companies of the Cameronians arrived, and it was essential to get them under cover at once. One company was placed in Hosack Trench, and the other two were sent back to the sunken road to occupy Hymen and the neighbouring trenches.

On the 14th the weather, which since the 9th had been abominable, improved. The Redoubt again came in for a good deal of shelling.

The Battalion was relieved after dark by the 1st Hampshires, and moved back into Brigade Reserve in Laurel Trench on the Brown line.

Thus ended the part played by the Battalion in the first phase of the Battle of Arras, 1917.

The first day was completely successful, but on the subsequent days the promise of a still greater success was not fulfilled. The Germans, on their part, made a quick recovery, and put up a skilful defence. It was a trying period: the weather was atrocious, cold with rain, sleet and snow.

Sleep was practically impossible, but all ranks showed a wonderful spirit. On one particular night there were only the three officers of Battalion H.Q. fit for duty, the remainder having temporarily collapsed from sheer exhaustion. Owing to the course of events it is impossible to give an accurate statement of the Battalion's captures on the 9th. As no escort was sent back, the prisoners were not counted; a low estimate is that at least 200, including several officers, were taken.

Similarly a large amount of war material, including many machine guns, was captured; but it was not possible to check it, or evacuate it from the battlefield.

During this period the Battalion's casualties were:—

Officers killed—

Captain H. J. Tanner.
Captain S. V. Wasbrough.
Sec.-Lieut. A. M. Hill.
Sec.-Lieut. C. M. Gardner.
Sec.-Lieut. S. H. Card.
Sec.-Lieut. E. R. Foy.
Sec.-Lieut. N. F. Herapath.

Died of wounds—
Sec.-Lieut. C. A. B. Elliott.

Wounded—
Lieut. C. J. O. Daubeny.
Sec.-Lieut. D. W. Gardner.
Sec.-Lieut. F. C. Barlow.

Other Ranks—127.

At 7 p.m. on April 16th the Battalion was relieved by the 2nd Seaforth Highlanders of the 10th Brigade in pouring rain, and moved back to the Black line (the old German front line system). Everybody was under cover of sorts, some in dug-outs, some in improvised shelters, and the majority under bivouac sheets: a much-needed rest was obtained.

On the 19th a warning for another attack was received, but this later was cancelled.

The Battalion was relieved on April 20th by a battalion of the Royal Fusiliers of the 37th Division, and marched to Agnes-les-Duisans.

On the following day the Brigade marched into good billets at Izel-les-Hameau.

After a short time devoted to rest, bathing and re-organisation, training was started. As was invariably the case, it was necessary to train a large number of Lewis gunners.

A draft was received, but owing to the shortage of men, companies were reorganised into three platoons. After a few days, from which much benefit was derived, and during which good use was made of the Lignereuil training area, it was time to move back to the line.

On April 28th the Battalion moved to huts just South of Acq, on the 29th to Y huts on the Arras-St. Pol Road, and on the 30th to billets in Arras.

The 4th Division relieved the 34th in the line, the 10th Brigade being on the right just North of the Scarpe, the 12th on the left, with the 11th Brigade in reserve.

In Arras the whole Battalion was billeted close to the canal in one big shed, which contained and was surrounded by thousands of shells of all sizes and kinds. It was far from being an ideal billet, and as a matter of fact there was a large explosion at this very spot a few days afterwards.

On the morning of May 1st it was announced at a Conference at Brigade H.Q. that another attack was imminent, that the 11th Brigade would be in Divisional reserve, but that the Battalion would be attached to the 10th Brigade, and would attack on its right to capture Roeux and Roeux Wood.

While the Division had been out of the line, a good deal of heavy fighting had taken place, but the result was only a total advance of a few hundred yards.

In the afternoon officers went to 10th Brigade H.Q., in Lancer Lane, on the South side of the Scarpe, opposite Fampoux. Here further details of the attack were received, and it was possible to get a fairly good, though a distant, view of Roeux. It was impossible to visit the front line.

At 8.30 a.m. the next morning, May 2nd, the Battalion left Arras and marched to the Ravine, West of Athies, and just North of the Fampoux Road, where it bivouaced for the remainder of the day.

At 8 p.m. the Battalion moved into the front line, relieving the Household Battalion of the 10th Brigade from Mount Pleasant Wood to the Scarpe.

Battalion H.Q. was situated behind a bank North of the river. During the night part of Roeux village was burning, and Germans could be seen moving about as though carrying out a relief. They were heavily fired on.

Plan of Attack.

The attack was on a large scale, and extended for a considerable distance on either flank of the 4th Division, especially to the South. Owing to its reputed strength Roeux was not to be assaulted at the beginning of the attack, but was to be kept under a heavy bombardment. The right of the advance of the 10th Brigade led just clear of the Northern edge of the village, which it was hoped to outflank.

Twenty minutes after zero the Battalion was to advance, capture and clear Roeux village and wood, and join up with the 10th Brigade on the first objective, the Black line, running approximately North and South, just East of the village. Later the 10th Brigade, in conjunction with the troops on its left and those South of the river, was to advance to the second objective, the Blue line, and afterwards to a third objective, the Red line, on the Western outskirts of Plouvain, while the Battalion remained on the Black line.

The Battalion attacked with three companies: D Company was detailed as the carrying company, and was accommodated close to Battalion H.Q.

Light (A) Company on the left and B Company on the right were the assaulting troops, each with two platoons in the front line and two in support.

C Company was detailed for "mopping up" the village and wood: half assembled with the leading platoons of Light and B Companies and half with the supporting platoons.

THE ATTACK OF MAY 3RD, 1917.

The attack was launched at 3.45 a.m.

At 4.5 a.m. the Battalion left its trenches and went forward to the assault in excellent style. It was a very dark morning, and this, in addition to the smoke and dust of the barrage, made it very hard to see. This resulted in a considerable loss of direction; the 10th Brigade came too far to the South and crowded the Battalion towards the Scarpe.

So few of the officers with the attacking companies came through the action unharmed, that it was difficult to gather a connected story of the day.

B Company, soon after entering the wood, was held up by machine gun fire and made no progress.

Light Company, on the left, reached the outskirts of the village, but was then also stopped by machine gun fire and local counter-attacks. Eventually the Battalion was driven back to its original front line. The same fate befell the 10th Brigade.

The 12th Brigade on its left made some progress, but by the afternoon had to give up practically all the ground gained. The attack all along the line was a complete failure. Generally speaking, on this part of the front the lack of success appears to have been due very largely to the fact that the attack was made in the dark, resulting in a considerable loss of direction.

From the Battalion's point of view, the following points undoubtedly influenced the result :—

(*a*) Roeux Wood did not receive sufficient attention from our artillery, and it contained many machine guns.

(*b*) Owing to the darkness these machine guns could not be located, and thus it was impossible for the attacking infantry to deal adequately with them with the weapons at their disposal.

(*c*) The attack was carried out across unknown country, and no preliminary reconnaissance of the front line had been possible.

(*d*) The maps available showed the right of the assembly trenches of the Battalion as bending somewhat towards the South-East. In reality, as a reconnaissance would have shown, there was a sharp curve, and the trenches here ran almost parallel to the river, and almost at right angles to the objective. Failure to realise this fact led B Company on leaving the assembly trenches to advance towards the river and not directly towards its objective. The attack thus lost cohesion, and B Company fell behind the creeping barrage.

After the attack had broken down, the Battalion was reorganised in the trenches from which it had assaulted.

During the night some forward posts were established by the 10th Brigade, joining up with the left of the Battalion.

The Battalion's casualties were :—

Officers : Died of wounds—
Sec.-Lieut. W. E. Marler.

Missing, afterwards reported killed—
Captain C. C. Codner.

Missing, of whom nothing has been heard—
Sec.-Lieut. R. H. D. Bailey.

Wounded—
Sec.-Lieut. R. J. Middleton.
Sec.-Lieut. S. R. Foley.
Sec.-Lieut. H. G. S. Backhouse.
Sec.-Lieut. F. H. Davies.
Sec.-Lieut. E. H. C. Frith.

Other ranks—132.

Captain E. W. Marshall was reported wounded and missing, but he rejoined unhurt after dark, having spent the day in a shell hole a few yards from the enemy.

At 11.30 p.m. strong patrols were sent out along the bank just North of the river to find out if the wood and village were still held. The enemy was found in strength and no progress was possible.

The next day, May 4th, was fairly quiet. During the morning patrols reported that the Western end of the wood was unoccupied. Accordingly, in the afternoon a patrol of fifty men with two Lewis guns was sent out to locate the enemy's position and to gain ground if possible. The Germans were found in strength in the middle of the wood, so posts were established in close touch with them. During the night these posts were joined up with the front line, North of the wood, and this gave a jumping-off line parallel to the objective for the next attack.

After dark the Battalion was relieved by the Household Battalion, and returned to the old German fourth system, North of Fampoux, coming again under the orders of the 11th Brigade and being in Brigade reserve.

After a short period of rest, in which a good deal of salvage work was done, the Battalion on the 7th moved again into the front line of the left sub-sector, relieving the 1st Rifle Brigade in front of the station buildings, just North of the Chemical Works: Battalion H.Q. was in an old quarry. Work was at once started on assembly trenches for another attack. The Germans seemed somewhat nervous, and their artillery was active.

On the 9th a small portion of the Battalion's sector on the left was taken over by the 50th Brigade of the 17th Division. Before dawn on the 10th the front line, for a depth of 500 yards, was evacuated in order that our heavy artillery might shell the station buildings, the Chemical Works, and the Cemetery.

The Battalion withdrew to Cove and Crash trenches, with Battalion H.Q. in the Eastern outskirts of Fampoux.

During the day the heavy artillery carried out an accurate bombardment, which called forth a good deal of retaliation from the German guns.

The Companies had orders to re-occupy their old positions at 9 p.m.

At 8.30 p.m., however, the Germans put down a heavy barrage on the unoccupied trenches, the S.O.S. was sent up on the right, and our guns at once opened with a still heavier barrage. The position was somewhat difficult owing to the evacuation of the front line, and the position not being favourable for repelling an attack.

Patrols were pushed forward at once, and they discovered that the enemy was not attacking: the trenches were rapidly re-occupied, and all was quiet by 9.30 p.m.

It is probable that the Germans expected an evening attack by us after the day's bombardment, and put down their barrage as a precautionary measure.

Considering the heavy shelling, the Battalion was lucky in only having eight casualties.

Later in the night the Battalion was relieved by the 1st Rifle Brigade, which took up its positions in the assembly trenches.

At 2 a.m., on the 12th, after the relief, the Battalion side-stepped a little to the South to Crete Trench, with Battalion H.Q. in Crump Trench, being in Brigade reserve.

PLAN OF OPERATIONS.

A local attack by the 4th Division was ordered to capture the Station, the Chemical Works, the Château, and the Cemetery. These positions had been the scene of much hard fighting, and had already withstood several determined assaults.

In order to effect a surprise, the attack was timed for 7.30 p.m., and the attacking troops had to remain very quiet and unobserved by aircraft in their assembly trenches during the day.

The 11th Brigade, from left to right 1st Rifle Brigade, 1st East Lancashires and 1st Hampshires, the Battalion being in reserve, was to capture the Station, Chemical Works and the Château, and establish a line just East of these positions.

The 6th Dorsets, of the 17th Division, were co-operating on the left to form a defensive flank.

The 10th Brigade on the right was given the task of capturing the Cemetery.

On the following morning the 17th Division was to continue the attack; a portion of the Rifle Brigade, the Battalion and the East Lancashires were to push forward in conjunction with the 17th Division, and establish a line 400 yards East of that captured on the preceding evening.

At this time the Battalion, owing to its weakness, was organised as two Companies only.

THE ATTACK ON THE CHEMICAL WORKS, MAY 11TH & 12TH.

At 7.30 p.m. on May 11th, the attack started behind an excellent barrage, a concentration of all available guns being brought to bear on the enemy's positions, and was entirely successful, several hundred prisoners being taken. As soon as it was dark, the Battalion carried up stores and ammunition to form forward dumps, and then moved into Cawdor Trench.

At midnight it took over the new front line just South of the railway, relieving two Companies of the East Lancashires, which side-stepped to the South.

During the attack the Germans shelled heavily, chiefly back by the river and on the vacated trenches. Fortunately the night was fairly quiet, and this facilitated the assembly.

At 6.30 a.m. on May 12th, the second attack took place, and was equally successful, little opposition being met. The new line was quickly consolidated.

The German guns were exceedingly active, but chiefly on the Chemical Works and the trenches further back. The new line was hardly touched.

The Battalion only lost one officer, Sec.-Lieut. Pratt, wounded, and 12 other ranks.

At midday Battalion H.Q. moved up into the Chemical Works.

News was received at 5 p.m. that the Battalion would be relieved by three platoons of the 1/8th Argyle and Sutherland Highlanders of the 51st Division, the whole Brigade front being taken over by this Battalion.

Owing to the short notice, no preliminary arrangements could be made.

The relief started about 10.30 p.m., but did not proceed very smoothly owing to the platoons of the Argyles losing their way, and the heavy shelling which lasted all night.

The incoming battalion had rather heavy casualties.

The relief was finally finished at 3 a.m. on the 13th: the Chemical Works and communications were being heavily shelled, but in spite of this the Battalion was most fortunate and had no casualties.

A halt for breakfast and a little rest was made at the railway embankment, West of Athies, after which the Battalion marched into billets in Arras.

Its casualties since May 3rd had been 9 officers and 160 other ranks.

On the next morning, May 14th, it proceeded in busses to Gouy-en-Tournois, where comfortable billets were found.

On the 16th the Brigade was inspected and addressed by General Allenby, the Commander of the Third Army.

A period of rest for nearly a month was now enjoyed, and very useful training was carried out. A Brigade Boxing Tournament was held, in which Sergt. Johnson and Sergt. Morris won their weights for the Battalion.

Drafts, but not in very large numbers, were received.

On June 11th a reconnaissance of the front line was carried out, and on the following day the Battalion was moved by busses into Arras: on the next morning it marched to the railway embankment West of Athies.

At 8.45 p.m. the same evening the front line immediately South of the Scarpe and opposite Roeux was taken over, the left resting on the river and the right in touch with the 3rd Division. The approach trenches were in a bad condition, Lone Lane in particular being 1½ feet deep in water, but the front line was fairly good. (Map 3.)

Work was at once started on the trenches, and active patrolling was carried out to determine exactly the enemy's positions. The posts on the tow path, S.W. of Roeux, were pushed forward 300 yards.

An aggressive attitude was adopted, and the enemy was harassed day and night by sniping, rifle grenades and Lewis guns. In spite of this he remained fairly quiet, except for a certain amount of trench-mortaring.

Detachments of the 3/4th Queen's Regt. were attached for instruction in the trenches.

After a period of considerable activity, the Battalion was relieved by the 1st Rifle Brigade on June 19th, and moved back into Brigade reserve at the railway embankment.

It held the front line again from the 25th to the 27th, on which day it was relieved by the 1st King's Own of the 12th Brigade.

The Brigade moved back into Divisional reserve, the Battalion being accommodated in tents in Rifle Camp, just East of Blangy.

A period of comparatively quiet trench warfare now set in.

The 10th Brigade held the line immediately North of the Scarpe, on a one battalion front: reliefs were carried out within the Brigade, and the Brigade did not leave the line.

The 11th and 12th Brigades took it in turns to hold the line just South of the Scarpe, with two battalions in the front line. A Brigade was in the front line for sixteen days, and then went into Divisional reserve for sixteen days.

While in the line, each battalion did eight days in the front line and eight days in Brigade support, or reserve. Taking it on the whole, the time was a pleasant one, and often recalled in later days with happy memories. At one time the German trench mortars became rather a nuisance, but our guns dealt with them effectually. All known or suspected trench mortar emplacements were carefully recorded, and the German position was divided into sectors, which were numbered.

When the trench mortars became active, the infantry in the line telephoned back the number of the sector giving the trouble, whereupon a concentrated burst of artillery fire was at once put on the offending sector, usually with the desired effect. A very great deal of work was performed, and a good trench system came into being. A great quantity of good wire was put out, dug-outs were constructed, and the trenches revetted.

The Brigade in reserve had to find large working parties, the equivalent of a complete battalion in each twenty-four hours. Each battalion in turn took over all working parties for a period of four days, and devoted the remainder of the time to training.

On August 16th the Division took over a portion of the 12th Division's front to the South, its right then resting on the Twin Copses. The position here was rather peculiar. Owing to the marshy nature of the ground, a trench was impossible, and the line was held by a few isolated posts. Furthermore, there was *no* support or reserve trench behind this gap, and the first organised line of resistance was East Reserve, in front of Monchy-le-Preux.

During the summer there were good reasons for supposing that the Germans were meditating a retirement on a big scale on this part of the line in order to shorten their front and upset our plans. Detailed orders to meet such a contingency were issued, and everything was in readiness if such a move had been made.

A good deal of the training was devoted to exercises in connection with closely following up a retiring enemy, and though the occasion did not arise, much useful information and practice were obtained.

In the middle of August, Brigade Aquatic Sports were held in the Scarpe: the Battalion did well, winning six firsts and three seconds.

During the first week in September the 4th Division was relieved by the 15th Division. The Battalion handed over Rifle Camp to the 11th Argyle and Sutherland Highlanders on September 5th, and marched to Bienvillers-au-Bois.

An excellent draft of 194 joined here: the majority were lads of nineteen, who had had seven to eight months' training. They were exceedingly keen and well trained, and subsequently did very well on their first introduction to war.

Throughout the summer the Battalion had been very weak in numbers, and it had been impossible to retain the four-platoon organisation.

As soon as this draft arrived, the Companies at once resumed the normal organisation.

During this period of training the 4th Division had the extreme misfortune to lose its Commander, Major-General Sir William Lambton, in whom all ranks had the greatest confidence. His horse fell with him, and he received such severe injuries that he was unable to take any further active part in the war. He was succeeded by Major-General T. Matheson.

CHAPTER IV.

BELGIUM—THE ATTACK OF OCTOBER 4TH, 1917, NEAR POELCAPELLE—THE WINTER OF 1917-1918—SPRING TRAINING, 1918.

THE main British attack of the year had by this time started in Belgium, and the 4th Division was notified that it would be called upon shortly to take its place in the battle line.

Training with this end in view was at once started, due provision as usual being made for football and other sports.

The Germans had by now developed a carefully thought-out method of defence, which met with a considerable degree of success.

Their system was to organise the defence in great depth, to dispense with regular lines of trenches, which became untenable owing to our artillery fire, the free use of scattered concrete "pill-boxes," and immediate counter-attacks in force to regain the ground lost at the first onslaught.

This necessitated a considerable alteration in our mode of attack: to counter the German system of defence, the following methods were evolved by us:—

(*a*) A barrage in great depth.

(*b*) An elastic formation to deal with the German shell-hole positions—this took the form chiefly of a thin line of skirmishers followed by lines of sections in single file.

(*c*) Careful arrangements for "mopping up," to deal with isolated pill-boxes.

(*d*) Making good tactical features, as the advance progressed, with specially detailed garrisons to act as "points d'appuie" in the event of a determined counter-attack.

(*e*) The detailing of specified units as counter-attack troops ready to move forward and meet the expected counter-attack by the Germans.

Full benefit was made of the lessons learnt in the recent fighting, and training was carried out on these lines.

On September 18th the expected move to Belgium started, and the Brigade marched to the Mondicourt area, the Battalion being billeted in Grenas. On the following day the Battalion entrained for Houtpertre, one mile S.W. of Poperinghe, and arrived in camp near Proven at 4 a.m. on the 20th.

Training was at once re-commenced, and the platoon football competition, which had been a great success, was finished.

Officers were conveyed in busses to Dawson's Corner on the 24th, and from there reconnoitred the forward area. Major R. H. Waddy, who had been second in command since August, 1916, left to take over command of the 2/5th West Yorkshire Regt. in the 62nd Division: he was succeeded by Major J. F. C. Mordaunt.

On the 28th the Battalion moved to the neighbourhood of Elverdinghe, and was accommodated in Cariboo Camp.

The personnel detailed to remain out of action moved back to Dragon Camp.

The weather was fine, and there was a glorious moon: the German aeroplanes took full advantage of this, and showed great activity. Every night their bombing machines were over our lines, and they inflicted considerable casualties on the large number of our troops which were concentrated in the neighbourhood, though fortunately the Battalion itself did not suffer.

On October 1st a move further forward was made to Hull's Farm, close to the Canal Bank. The Battalion was now only a short distance from where the gas attack of August, 1916, had been experienced. The line by this time had been considerably advanced by the attack then in progress.

PLAN OF OPERATIONS FOR THE ATTACK OF OCT. 4TH, 1917.

(Map 4.)

The XIV. Corps, with the 4th Division on its right and the 29th Division on the left, and the XVIII. Corps, with the 11th Division on its left, were to attack and capture the Green line, or final objective.

It was arranged that a pause of one hour should be made on the first objective.

The 4th Division attacked with the 11th Brigade on the right, the 10th Brigade on the left, and the 12th Brigade in reserve.

The 11th Brigade attacked with two battalions in the front line, one in support and one in reserve.

The Somerset Light Infantry was the right and the Hampshire Regt. the left assaulting battalions: the East Lancashires were in support and the Rifle Brigade in reserve.

Battalion arrangements. As a result of the German system of defence, previously mentioned, careful precautions were taken to ensure that "mopping up" was thoroughly carried out, and to guard against counter-attacks. These are given in detail, as the orders for this action were far more complicated than for any other.

The portion of the Green line allotted to the Battalion as its objective was from V19b28 to V13c35.40. Its boundaries on either flank are shown on the map.

The Battalion attacked on a two-company front, Light Company on the right, B Company on the left, with H Company in support on the right and C Company in support on the left. The leading companies were ordered to go right through to the final objective.

The dividing line between companies was a line through U24c60.45—U24d17.85—U24b74.22—V19a48.70.

Moppers up. H and C Companies each detailed two platoons as moppers. The leading platoon of H Company followed closely behind the leading two platoons of Light Company, and was ordered to mop and hold Kangaroo Trench. The second platoon of H Company, following close behind its first platoon, was ordered to take, mop and hold the concrete pill-box and vicinity in U24d60.95.

Similarly the leading platoon of C Company, following the leading two platoons of B Company, secured its portion of Kangaroo Trench, while the other platoon was made responsible for the strong point and vicinity at U24b73.

H Company also detailed four sections, two to follow the right rear and two the left rear platoon of Light Company.

Similarly C Company detailed four sections, two behind each of the rear platoons of B Company.

These sections were responsible for mopping up the various pill-boxes and strong points captured by the platoons which they were following.

These platoons had allotted to them strong points as follows:—

Light Company—Right rear platoon, Ferdan House.
Left rear platoon, Pill-box V19a05.27.

B Company—Right rear platoon, Strong point 47 West of Kangaroo Pond.
Left rear platoon, Lemnos House.

After capturing these points these platoons were ordered to hand them over to their attendant mopping sections, and push on with the least possible delay after the leading troops.

It will thus be seen that each of the support companies had to find three platoons for mopping up, and that the force available to capture the final objective was not large. The method of attack adopted was somewhat involved, and is open to criticism, but in view of the length of the advance and the obstacles encountered, its advantages were considered to outweigh the disadvantages, and it proved successful. Its chief advantages were that definite parties were detailed to mop all known or suspected points of enemy resistance, and that as the advance proceeded formed bodies were dropped, which formed "points d'appuie" in case of an enemy counter-attack, and the Battalion was invariably distributed in depth.

The leading troops were concerned solely with following close behind the barrage, dealing only with such Germans as actually obstructed their advance, and leaving the garrisons of all pill-boxes and strong points to be rounded up by the various parties detailed for the purpose.

After the objective had been captured, the rearmost parties, on the completion of their particular duty, moved forward and consolidated with the remainder of the support companies in depth.

Two machine guns and one Lewis gun of the 21st West Yorkshire Pioneers were attached to the support companies: the Lewis gun was for anti-aircraft work, and with a view to economising ammunition all other Lewis guns were forbidden to fire at hostile aeroplanes.

One platoon for each of the support companies was detailed by name for counter-attack purposes.

Two platoons of the 1st East Lancashire Regiment, with two machine guns, were attached to the Battalion to act as a flank guard, assist the 33rd Brigade on the right, and fill any gap that might occur between the two Divisions.

The creeping barrage fell 150 yards in front of the front line, advancing at zero—3 minutes at the rate of 50 yards in two minutes for the first 200 yards, and then at the rate of 50 yards in three minutes up to the first objective: from there to the final objective it moved at 50 yards in four minutes.

A machine gun barrage of eight guns was also put down on the line V13a63—U18b95 from zero to zero + 30 minutes.

The weather was now bad, heavy rain had fallen, the shell holes were full of water, and the mud was abominable.

Consequently to avoid exhausting the men, it was decided to assemble the night before the attack and not earlier. As zero was timed for dawn, this gave the assaulting troops no opportunity of studying by daylight the ground over which they had to attack. To minimise as far as possible this disadvantage, 6 officers, 8 N.C.O.'s and 8 men were sent into the front line on the evening of the 2nd. They were thus able on the 3rd to study the ground, take compass bearings, and also lay out tapes after dusk, on which the Battalion had to assemble in the absence of trenches.

At 5.30 p.m. on the 3rd, the Battalion left Hull's Farm and marched up to the assembly area in the front line.

Owing to the bad state of the ground duck-board tracks had been laid almost to the front line, and this greatly facilitated movement. These tracks, however, were well known to the enemy, and usually were continuously shelled by the German guns, which were in great strength.

Most fortunately the night was exceptionally quiet, and hardly a shell fell during the march forward. The Battalion was in position by 11 p.m., with the loss of only one man.

It assembled in three lines: the first, consisting of the leading platoons of Light and B Companies with the mopping platoons of C and H, on the front line of posts running approximately from just in front of Louis Farm to U24c95 35. The second, consisting of the support platoons of the leading companies with the mopping sections of C and H, on the line of the hedge running S.E. from Eagle House; and the third, consisting of Company Headquarters, the fourth platoons of C and H, and the attached platoons of the East Lancashires, on the line of the hedge running S.E. from the concrete pill-box in

U24c27. Battalion H.Q. was established in the pill-box at U29b65.95.

The Attack of Oct. 4th, 1917, at Poelcapelle.

At 6 a.m. the attack started. The men were imbued with the idea of keeping close to our barrage, and in their eagerness were inclined to keep almost too close, with the result that a good many casualties were caused at the beginning by our shells. Owing to the darkness it was hard to keep direction, and some confusion ensued soon after the start, the tendency being to ease off to the right.

Germans were found in Kangaroo Trench, but they gave little trouble, and our barrage had accounted for a good many.

No serious opposition was met until the line of the track running South from Lemnos House was reached. The Germans in some force were holding piles of stones on the road about V19a25.20, but they were rapidly dealt with by fire and several killed.

The right of the Battalion, together with men of the 11th Division, then advanced against the concrete house at V19a60.15, which was in the 11th Division's area. One machine gun was firing from the top of the house and another from the side. A Lewis gun was brought into action, and under cover of its fire our men worked round a flank and got in rear. Several Germans were killed here, and sixteen prisoners taken, together with two machine guns.

On the left a machine gun, firing from Lemnos House, held up the attack for a time, but it was knocked out eventually by rifle fire. Machine gun fire from Ferdan House was also troublesome.

The barrage then halted for an hour in front of the first objective; the men were reorganised as far as possible and the direction was checked. The opportunity was taken of cleaning Lewis guns and rifles, all of which were covered with mud.

When the barrage moved forward the advance continued, the right directing on Ferdan House, which was found to be strongly held. Lewis guns were brought into action against it, and a Stokes mortar of the 11th Division gave valuable assistance. Under cover of this fire the flank guard of the East Lancashires worked round the right flank, while men of the Battalion went round the left. Two officers and thirty men, many of them wounded, were captured in addition to two machine guns and two trench mortars.

The advance from here to the final objective met with little opposition, though machine gun fire from the direction of 19 Metre Hill caused some casualties.

The Battalion established a line of posts on the Green line from Tragique Farm to about V19b18, and touch was established on both flanks.

Only five officers and about the equivalent of four platoons of the Battalion were left, with about thirty men of the East Lancashires under an officer.

Two platoons and the party of the East Lancashires held the front line, with the remainder in support.

During the attack a very gallant deed was performed by Lance-Corpl. Watkins, of C Company. A portion of the line was held up by fire from a pill-box. This gallant N.C.O., though already wounded in the arm, went forward alone under heavy fire, worked round a flank, and attacked the pill-box single-handed, capturing the machine gun and its crew, and enabling the remainder of the line to advance. He was rewarded with the D.C.M., but unfortunately, after having been promoted Sergeant, was killed by a chance shell the following summer.

About 2 p.m. a counter-attack took place on the left, on 19 Metre Hill, and the Seaforths on the right of the 10th Brigade were seen to be falling back. The left of the 11th Brigade had to give some ground also in order to conform.

The Battalion inflicted heavy losses on the enemy advancing over the ridge with Lewis gun, machine gun and rifle fire. This assisted considerably the troops on the left to recover the lost ground, and the line soon afterwards was re-established.

The Battalion lost a good many men from German rifle fire.

At about 5.40 p.m. another counter-attack was launched on the left, and the line was forced back, but the Battalion again held its ground.

Shortly afterwards, as our guns were persistently shelling the Green line, the front posts were withdrawn to a line 100 yards in front of Lemnos and Ferdan Houses in order to avoid casualties.

A satisfactory feature of the day was the way in which the last big draft behaved: though for the most part only 19 years of age, and never having been under fire before, they showed the greatest keenness and determination.

In this action the Battalion lost 9 officers and 282 other ranks.

Officer casualties:—

Missing, of whom nothing further has been heard—

Sec.-Lieut. A. R. Barnes.

Wounded—

Captain D. C. H. Edwards.
Captain C. A. S. Hawker.
Lieut. S. V. Butcher.
Sec.-Lieut. C. E. Matthews.
Sec.-Lieut. P. J. Sylvester.
Sec.-Lieut. L. H. Mogg.
Sec.-Lieut. J. W. Harper.
Sec.-Lieut. E. G. Hoskins.

The following day, October 5th, was quiet in the morning, but heavy shelling took place in the afternoon.

After dark the Battalion was relieved by the 1st Rifle Brigade, and moved back to Candle Trench.

The weather had been wet and stormy, and the condition of the ground became even worse than before.

Owing to this it was impossible to bring back the numerous machine guns and trench mortars that had been captured.

On the 7th the Battalion went into Divisional reserve in Leipzig Farm.

The Division continued the attack on the 9th, and the Battalion "stood to" at zero, ready to move at a moment's notice if required. The occasion, however, did not arise.

The next few days were devoted to refitting and making the camp habitable: the weather still remained wet.

On the 12th the Battalion marched to Elverdinghe, entrained for Proven, and proceeded to Paddington Camp. The accommodation was good, but scattered, and the surroundings very muddy.

Brigadier-General R. A. Berners, D.S.O., was transferred from the Division, and handed over command of the 11th Brigade to Brigadier-General T. Wade, D.S.O.

On the 15th a move was made to Poperinghe, where the Battalion, though rather crowded, found comfortable billets. A draft of 65 men joined here.

The 4th Division now returned to the Third Army.

The Battalion entrained at Pezelhoek at 10.30 a.m. on the 18th, detrained at Marceuil at midnight, and arrived in billets in Dainville at 2.30 a.m. on the 19th.

On October 22nd it marched to the Bois des Boeufs Camp, where it was in Divisional reserve to the Monchy sector, and on the following night relieved the 8th Royal Fusiliers of the 12th Division in the front line. The left then rested on the gap in the Twin Copses previously mentioned, the Battalion being immediately South of the sector held previous to its visit to the Ypres battlefront.

The trenches handed over by the 12th Division were distinctly good, and showed signs of much hard work, especially taking into consideration that it was weak in numbers and had been carrying out constant raids.

Heavy rain fell, and this damaged the trenches considerably. Revetting with wooden frames was at once taken in hand, and the systematic wiring of the front undertaken.

From now onwards an excellent system came into being. The whole work on the Divisional front was co-ordinated and controlled by the C.R.E. The infantry holding the line supplied working parties under R.E. supervision; this ensured continuity of policy and avoided delay when reliefs took place. The results were very satisfactory, as were proved by the amount of work carried out.

A period of ordinary trench warfare during the winter months now set in. Elaborate precautions for the health and comfort of the men were taken. The trenches for the most part were good, revetted and duck-boarded, and a considerable number of dug-outs existed.

Full benefit was made of the experience gained in previous winters: drying rooms were constructed for gum-boots, which were provided on an adequate scale and a regular use of whale oil was instituted to prevent trench feet. Arrangements were also made to supply hot meals to the men in the front line. The results were excellent, and the health of the Division remained very good.

The Division held the line until the first week of February, 1918. During this period the Battalion held various parts of the front, from the Twin Copses as far South as the Arras-Cambrai Road.

The usual tour in the front line was four days, with similar periods in Brigade support or reserve. Each Brigade had, as a rule, twelve days in Divisional reserve, which were spent either in the Schramm Barracks in Arras, or in the Bois des Boeufs or Wilderness Camps on the Cambrai road.

In view of the expected German offensive in the spring, the work of strengthening the defences was pushed on with the utmost energy. This entailed a great deal of labour, and battalions even when out of the line and resting had invariably to find large working parties.

Reference has previously been made to the gap in the defensive line which existed West and South-West of the Twin Copses owing to the marshy ground. In spite of the many unsuccessful attempts which had already been made, Divisional H.Q. decided to dig a support line here, and Dale Trench was the result. By skilful construction, and a thorough system of pumping, it proved a success, and was kept clear of water.

This trench was strongly wired with barbed concertina; the greater part of this wire was made by the Battalion when in Brigade reserve in trenches just East of Feuchy Chapel cross-roads, and carried up to Dale Trench. The actual erection of the wire was done entirely by the Battalion, and an excellent job was made of it.

This winter in the line was on the whole uneventful: the artillery on both sides showed considerably activity, and gas shells were freely used. The German trench mortars were, as usual, rather a source of annoyance, but as a general rule they were kept in subjection by our guns. At the end of November certain "quiet periods" were ordered; during these periods no gun was permitted to fire except in case of emergency. The idea apparently was to mystify the enemy as to our intentions; it is not known if the desired result was obtained, but the Germans invariably replied to such "quiet periods" by particularly heavy shelling, and the infantry were distinctly adverse to the procedure.

During the first half of December an enemy attack was expected: from the 10th to the 18th the Battalion, which was then in Divisional reserve in Schramm Barracks, Arras, "stood to" every morning at 6.30 a.m. ready to move at a moment's notice. No such attack developed, and there is reason to suppose that the Germans were suffering from similar apprehensions and were taking the like precautions.

A good deal of rain fell at the end of October, but November on the whole was fairly fine. December was cold, with a continuous frost and some snow. This weather continued well into January, but a sudden change occurred on the 15th, and heavy rain fell. This on top of the hard frosts and snow did great damage to the trenches, and much to the disappointment and surprise of everybody after all the hard work that had been performed, the majority proved unequal to the strain and fell in. The conditions then in a minor degree became similar to those on the Somme in the preceding winter; the communication trenches were quite impassable and duck-board tracks in the open had to be laid to the front line.

On Christmas Day the Battalion was in Brigade support, with H.Q. and one company in Fosse Farm, one company holding C, D and E strong points, and the other two companies in Fork Trench and East Reserve. In the evening it moved into the front line on the Cambrai road.

The Christmas celebrations were held in Wilderness Camp on January 5th.

On January 31st, 1918, Brigades were reduced from four battalions to three, and the 1st East Lancashire Regt. was transferred to another Division.

Early in the year the A.R.A. Competition for platoon field firing, combined with the bayonet assault, was started. The intention was that eliminating competitions should be held within battalions, brigades, divisions and corps, with the idea of eventually selecting the best platoon in the Army in France. Owing, however, to the German offensive, the competition could not be carried through to its final conclusion.

On February 2nd the Battalion competition was held on Wailly Range, and No. 6 Platoon of B Company provided the winning team.

In the first half of February the 4th Division was relieved by the 15th. The Battalion handed over Wilderness Camp on the 5th to the 11th Argyle and Sutherland Highlanders, and marched to Wanquetin.

The accommodation was in huts, and not too good. In spite of this a pleasant six weeks' rest was much enjoyed by the men. A good training area was available, and most successful training was carried out, the value of which was fully apparent later in the year.

By this time it was clear that the Germans were preparing an offensive on a very big scale, the only points in doubt being the date and the exact locality.

The Division being in Corps (XVII.) reserve, had to be prepared to play many rôles: the front on which it might have to operate extended from the Point du Jour, North of the River Scarpe, as far South as Mercatel.

No less than nine different defence schemes were drawn up to meet the various contingencies which might arise. This entailed much hard work for the officers: in addition to the training, which never ceased, these schemes had to be studied and elaborated, and many reconnaissances of forward positions had to be carried out.

In view of the expected German attack, advanced training was rapidly pushed on, and little time was allotted to platoon training. This, though understandable, was in many ways regrettable, and prevented the greatest value being obtained. As the winter, following immediately after the attack of October 4th, had been spent in the line, there had been little or no opportunity for training, and the general knowledge within the Battalion of open warfare, with which the whole of the training was concerned, was distinctly deficient.

The time for preliminary training being so restricted, it was only possible to concentrate on a few of the most important points in order to take part in battalion and brigade field days, which started almost immediately.

The principles of fire and movement were, however, thoroughly grasped, and the value of this training was fully proved in the later stages of the war.

In addition to the various phases of open warfare, gas drill and gas discipline received great attention, owing to the ever increasing use of gas shells by the enemy.

The march back to camp from the training area was usually carried out with box respirators adjusted: not only was great speed in getting protection obtained, but the box respirator became an integral part of the man, so much so, that cases are on record of men sleeping peacefully with their box respirators on.

Sports of all kinds took place, and a Light Company Competition was held, which was won by H Company. A most successful Boxing Tournament was held, for which over 200 entries were received: the men entered and boxed to gain a point for their Company, and some very spirited bouts took place.

The Battalion turned out a good Rugby football team: a Divisional Competition was held on a Brigade basis, and this was won by the 11th Brigade, eleven of the team being provided by the Battalion.

Progress was made with the A.R.A. Competition, and No. 6 Platoon won both the Brigade and Divisional events. Subsequently it was found impossible to carry the competition into formations higher than a Division.

Early in February the Battalion discarded the yellow ribbon, and started to wear a piece of Regimental ribbon on both arms as a distinguishing patch, and this was continued until the end of the war.

The German offensive was expected during the second week in March, and the Battalion was in a constant state of readiness to move.

The Division was due to relieve the Guards Division, North of the River Scarpe, but owing to the expected attack the relief was postponed.

CHAPTER V.

THE GERMAN ATTACK ON ARRAS, MARCH 28TH, 1918—LA BASSÉE CANAL—CAPTURE OF RIEZ DU VINAGE, APRIL 14TH, 1918—SUMMER OF 1918.

THE German attack, however, did not take place as soon as was expected, and it was decided to carry out the Divisional relief. (Map 3.)

On March 19th the Battalion left Wanquetin and moved to Schramm Barracks in Arras.

On the following evening it relieved the 1st Irish Guards in the front line of the centre sector, North of the Scarpe. The sub-sector held was that immediately North of the railway, with Battalion H.Q. in Cadiz Trench.

The Battalion was conveyed from Arras to Fampoux by railway: the detrainment in the centre of Fampoux was rather an anxious time, as the village was constantly heavily shelled, but the night was quiet and there was no difficulty.

On the next morning, March 21st, the German offensive started: no attack took place on the Divisional front, but throughout the day and night very heavy shelling was experienced, though fortunately few casualties were incurred.

The next day was comparatively quiet, but disquieting news of the German advance on our right was received. To conform to the line further South, Monchy was evacuated, and this was carried out unmolested by the Germans.

The right of the 4th Division, resting on the North bank of the Scarpe, was now considerably exposed.

The loss of Monchy without a fight, on the defences of which so much labour had been expended during the preceding winter, and in the strength of which everybody had the greatest confidence, was keenly felt by the whole of the 4th Division.

From dawn on the 23rd the front line on the Divisional front was evacuated by day, but was kept under observation by patrols. During the afternoon the Germans became inquisitive, and sent out parties to find out the situation. These were successfully dealt with by the Battalion's patrols, and only one party of Germans managed to enter our trenches. They were, however, quickly discovered, and at once ejected.

On the 24th, after a quiet day, the Battalion handed over the front line to the 1st Hampshires, and moved back into Brigade reserve, Battalion H.Q. with three companies being in Stirling Camp, on the Western side of the railway embankment, West of Athies, and B Company being left in Pudding Trench, just West of Fampoux.

On March 25th A Company was sent forward to join B in Pudding Trench.

Orders were received at night to work on the Army Line which ran just on the Eastern side of the railway embankment. This line was rapidly put in a state of defence, and dumps of ammunition were formed in it. It was manned from 5 a.m. till 10 a.m. on the 26th.

The Divisional Pioneers, the 21st West Yorkshire Regt., had arrived in camp, and came under the orders of the Commanding Officer for the purpose of holding the Army Line.

During the night of the 26-27th March, and the early morning of the 27th, heavy shelling took place, and orders were received for the half battalion in Stirling Camp to move forward into Pudding Trench. This order was cancelled shortly afterwards, as nothing serious had occurred, and only a raid on the left of the Divisional front had been attempted.

At 3 a.m., on March 28th, a heavy barrage fell North and South of the Scarpe, together with a considerable quantity of gas.

About 6 a.m. orders were received to move the two companies in Stirling Camp forward to join the remainder of the Battalion in the Third Trench system. Shortly afterwards these two companies and Battalion H.Q. marched off. The Fampoux road was being heavily shelled, and there was so much gas about that box respirators had to be worn. A track North of the road, however, was used, and the companies arrived in position with the loss of only one man: Battalion H.Q. was established in Pudding Trench.

At 7.30 a.m. the German attack was launched. A magnificent defence was offered by the 4th Division and the 56th Division on its left: the enemy were held, and were only able to occupy our most advanced trenches. The attack was made in great depth with large forces. Greenland Hill was alive with infantry in massed formations, advancing to exploit the success expected from the leading troops, with guns, limbers, wagons, and mounted officers.

Wonderful targets were offered to our artillery, but unfortunately communications generally failed owing to the enemy's bombardment, and in many cases it was impossible to get the guns on to the targets.

Excellent work was performed by one or two artillery forward observation officers, who in spite of every difficulty managed to keep their lines going through most of the day, and they had some magnificent shooting.

At 9.30 a.m., owing to the pressure of the enemy to the South, C Company was sent to form a defensive flank facing South along the Scarpe from the front line Third system just West of Fampoux to the Army Line, and to guard in particular all the bridges.

At noon the Hampshires were holding Cadiz and Coot trenches: the 12th Brigade on the left prolonged this line to the North, but it was necessary to send forward one company of the Rifle Brigade, who were in support, to fill a gap between the two brigades.

The 10th Brigade on the right also broke the enemy's attack, but as its right flank was very much exposed owing to the German advance South of the river, it had to withdraw about 2 p.m., but through a misunderstanding of the orders received it retired too far, almost to Athies.

The Hampshires still held on to the Cadiz line, with their right on the railway.

No further change took place during the remainder of the day.

At night a re-adjustment of the line became necessary as a result of the day's fighting, especially owing to the German advance South of the Scarpe, which had almost reached the Eastern outskirts of Feuchy.

The Hampshires were withdrawn to the trench running from Effie Trench, through H9b & c to the East of Athies.

The Rifle Brigade held Stoke Trench as the new front line running through Fampoux from Camel Avenue to the Scarpe.

The Battalion stood fast, the only change being that Light (H) Company moved back to Carolina Trench, the reserve line of the third system.

During the 28th the Battalion had not been involved in the actual fight, but had been subjected to heavy shelling, and was fortunate in only having twelve casualties.

On the 29th the Germans started heavy shelling at an early hour. About 7 a.m. it was reported that Germans were in Stoke Trench, and were bombing down Camel.

O.C. A Company, Captain Parsons, whose company was holding Port and Pudding Trenches, was ordered to assist the Rifle Brigade by sending a bombing party down Camel Avenue. As this did not prove successful, A Company was ordered to counter-attack at once across the open from Port Trench with two platoons and retake Stoke Trench.

This was carried out most successfully, and in spite of considerable machine gun fire Stoke was recaptured and the Germans ejected.

This operation, though a minor one, reflected great credit on all who took part in it. Immediate action was imperative, and there was no time for detailed orders. Lieut. E. M. Prince, who was in command of the counter-attack, only had time to pass word down the line that the two platoons concerned would attack at once, and that the signal to start would be a Very light. The men responded right nobly: on the signal being given, every man was up and over the parapet.

The casualties were heavy in proportion to the numbers engaged, and consisted of one officer, Sec.-Lieut. K. L. King-King, wounded and 27 other ranks. Captain A. C. Parsons was shortly afterwards killed while trying to bring in the wounded.

These two platoons remained as the garrison of Stoke Trench until relieved later by the Rifle Brigade. Their place in Port Trench was taken by two platoons of Light Company, which moved up from Carolina Trench.

During the afternoon Pudding Trench was heavily shelled.

The situation now was that the Germans on Orange Hill were able to overlook all our trenches on the rising ground North of the river. Similarly we were able to obtain direct enfilade observation of the German positions, though the advantage on the whole rested with the enemy owing to the configuration of the ground.

The result was that the infantry on both sides came under accurately observed shell-fire, and had a far from comfortable time.

The Battalion's casualties to date were three officers—Captain A. C. Parsons killed, Sec.-Lieuts. H. L. King-King and F. W. Perrett wounded—and 80 other ranks.

On March 30th Pudding and Port, and, in a lesser degree, Dingwall Trenches, were heavily shelled in the morning, and again in the afternoon. These trenches were far from good to start with, and were rapidly showing signs of their rough treatment.

Captain A. H. Collins was wounded.

During the night A Company relieved C, and took over the defensive flank along the river.

Pudding and Dingwall Trenches again came in for heavy shelling on the morning of March 31st, mainly from batteries that had moved up behind Monchy.

On April 1st the Battalion, on relief by the Hampshires, moved forward and relieved the Rifle Brigade in the front line.

Three companies—Light, A and B—held the front line Stoke Trench from 150 yards North of Camel Avenue to the Scarpe; C Company was in support in Pudding and Port. Rain now fell, and the artillery fire continued somewhat active

On April 5th the enemy carried out a heavy and systematic bombardment: our casualties were light, but much damage was done to the trenches.

The Battalion was relieved on the night of April 5th-6th by the Hampshire Regt., and moved back into Brigade reserve in and just to the North of Athies.

The Division was now relieved by a Canadian Division, and on April 7th, Jellalabad Day, the Battalion was relieved by the 10th Canadian Battalion, marched to St. Laurent Blangy, and from there was conveyed in busses to Y hutments North of the Arras-St. Pol road, close to Duisans.

On the afternoon of April 9th the camp was persistently shelled by high velocity guns, and had to be vacated: in the evening the Battalion marched to Hermaville.

A welcome draft of 3 officers and 162 men joined here.

Orders were issued that everything had to be in readiness for a move to the North at three hours' notice.

At 8 a.m., on the 12th, the orders to move were received, and at 11 a.m. the Battalion marched out of Hermaville and embussed at 1 p.m. with the rest of the Brigade at

Duisans. The convoy proceeded North *via* Houdain and Bruay, and the Brigade debussed just outside Lillers, on the Busnes road, at 6 p.m. (Map 5.)

On arrival it was discovered that the situation was very obscure, and no definite information could be obtained. It was known that the Germans had attacked in force, had overwhelmed the Portuguese, and had made a considerable advance, but the exact position of the front line could not be ascertained.

The 11th Brigade, covered by the Hampshires as an advanced guard, moved forward at 7 p.m. to take up a position on the La Basseé Canal. This proved to be the front line, but it was exceedingly lightly held by detachments of various Divisions and formations. Men of the 51st Division, which had been sent forward in support of the Portuguese, and had had heavy fighting, were here, as well as units of the 3rd Division. In addition there were men of King Edward's Horse, and men of the Tank Corps on foot with Lewis guns.

The Battalion took up an outpost line on the La Basseé Canal, with C and B Companies holding from Pont Levis, at Douce Crême Farm, P36b28, to Pont Levis, at Q32c68. Light Company was in support in Mont Bernenchon, and A Company in reserve in Bellerive: Battalion H.Q. was in a farm at V5b50.

The next day was quiet: it was evident that the German advance was at a standstill for the time being, and that the enemy's infantry had got well ahead of his guns. The surroundings were completely novel: the countryside was green and untouched by war: the farms were standing and were full of furniture, food and live stock.

This change from the dismal landscape of previous trench positions was very marked and very welcome in most ways. The tragedy of the situation, however, was vividly brought to mind by the streams of refugees, who had been met on the march forward, fleeing before the Germans.

The only trenches that existed were on the South bank

of the Canal: these had been constructed long before by the French, but proved of great value.

Neither side knew with any certainty the dispositions of the other, though during the day our patrols established the fact that the enemy were holding the village of Riez du Vinage.

The 61st Division, on our left, were holding the vicinity of Carvin.

This day presented some unusual sights: the British front line ran along the South bank of the Canal, but in spite of this the tow-path was alive with French civilians on bicycles and even in carts, busily intent on rescuing their most cherished possessions from their deserted homes.

The German guns were presumably moving forward to their new positions, and this unusual movement in the front line drew no hostile shelling.

Before steps were taken to stop it, quite a number of French civilians crossed the Canal and proceeded to their homes in no man's land. The fate of these in most cases is not known, but unfortunately many lost their lives or were wounded in the subsequent operations.

At noon, on April 14th, a Conference of Commanding Officers was held at Brigade H.Q. It was announced that it was imperative to establish a bridge-head of some depth North of La Basseé Canal, as this Canal, which formed a considerable obstacle, was the only natural line of defence which lay between the Germans and the coal fields around Bruay. The necessity of securing this bridge-head was the factor which governed the subsequent operations in this part of the front.

As the first step, Riez du Vinage was to be captured at once.

The Battalion was selected for the attack, which was ordered for the same evening at 6.30 p.m.

Plan of Operations.

The objective ran from the enclosure at Q26a50 along the Northern outskirts of the village to the road at Q26b72, through Q26d82, and thence back to the Canal at Q32d38.

The 1st Hampshire Regt. was ordered to move forward on the left, but in rear, and join up from the left of the Battalion with the 61st Division in Carvin.

The objective was allotted to companies as follows :—

C Company from the enclosure at Q26a50 to the road at Q26b72.

B Company from the road Q26b72 to Q26d82.

Light Company from Q26d82 to the Canal at Q32d38.

A Company was in reserve.

The Battalion was holding the Southern bank of the Canal, though a few advanced posts had been established the previous night a short way along the road to Riez. Owing to the few bridges available the Battalion had to assemble in daylight the other side of the Canal before attacking.

Assembly areas were allotted as follows :—

Light Company, the houses just North of the Canal in Q32a60.

B Company, the houses along the Riez road, held by the advanced posts, and houses on the North bank of the Canal, close to the footbridge, at Q31a82.

C Company, the houses North of the Canal, about B in La Basseé Canal.

A Company, the houses North of the Canal close to the drawbridge at P36a88.

In order not to arouse the enemy's suspicions, all movement forward to the position of assembly was by ones and twos. Orders were issued that no movement of formed bodies would take place before 6 p.m., at which hour the heavy artillery would start to bombard Riez.

At 6.30 p.m. the 18pr. barrage was due to fall on the Western edge of Riez, starting to move forward through the village at 6.40 p.m.

The line of assembly was the North bank of the Canal, and in order to get parallel with the objective it was necessary for the Battalion to deploy facing half-right before zero. The line selected for deployment, and from which to launch the attack at 6.30 p.m., was a North-West and South-East line through the fork roads at Q26c20, with C Company North and B Company South of the road, each Company on a 300 yards' frontage; Light Company with its right on the drawbridge in Q32c, extending for 300 yards along the road to Riez, and A Company facing North-East, with its centre on the road about Q31b58.

At 6 p.m. Battalion H.Q. moved to a house North of the Canal, just over the footbridge in Q31a.

Arrangements were made for machine guns to barrage the West side of the Bois de Pacaut during the attack.

The Conference at Brigade H.Q. was held at 12 noon, and by 2 p.m. Company Commanders had received verbal orders at Battalion H.Q.

By 6 p.m. the whole Battalion had left its positions South of the Canal, had dribbled across the Canal in twos and threes, and was assembled North of the Canal.

This dribbling was a lengthy proceeding; it will thus be seen that there was very little time for detailed instructions or explanation to the men of what was required of them. The operation presented considerable difficulties. The dribbling across the Canal was not easy, and the men had to act largely on their own initiative.

Then a deployment facing half-right had to be made in order to get into position for the attack, and finally the objective was almost semi-circular in shape. Consequently the fact that this attack, which was carried out by the Battalion alone, was so successful reflected great credit on all ranks, and proved the value of the Wanquetin training.

The Attack of April 14th, 1918, on Riez du Vinage.

At 6.10 p.m. the Battalion started to move forward to its position of deployment, and was ready to advance in artillery formation at 6.30 p.m.; it was however some distance short of the line from which it had been decided that the attack should be launched; in consequence the leading troops were unable to get close up to the barrage.

Shortly before zero some heavy shells fell North of the Canal in Q31, and caused several casualties.

At 6.30 p.m., when the creeping barrage started, machine gun fire was opened from Riez and the Bois de Pacaut. Owing to the accuracy of this fire the advance was temporarily held up about Q3.c6.9, and no progress could be made. Captain L. A. Osborne, M.C., in command of A Company in support, seeing that the situation was critical, and that the advance had to be continued at once if the attack was to be successful, immediately led forward two of his platoons through the leading companies, inspiring them by his personal example. The lead thus given by A Company was quickly followed by the remainder of the Battalion, and the whole line advanced by section rushes, the two platoons of A Company being the first to reach the village.

Light Company in the meantime on the right had also come under heavy machine gun fire from the Bois de Pacaut, and the two left platoons were checked. The other two platoons continued to advance by section rushes. In the end the machine guns in the Bois de Pacaut were silenced by our artillery and machine gun barrage, and Light Company took up its allotted position.

As our leading troops reached the village some Germans ran out and surrendered: determined fighting, however, took place in the village, chiefly in the Southern half, and a good number of Germans were killed.

When our men were approaching the Eastern outskirts, the Germans delivered a counter-attack, consisting of at least one company with six machine guns, from the direction of Q27 central and the Northern end of the Bois de Pacaut.

This counter-attack was at once engaged by rifle and Lewis gun fire, with excellent results. It was estimated that casualties to the extent of 40% were inflicted on the enemy: half the remainder ran away, and the other half ran forward with their hands up and surrendered. Subsequently the six machine guns were collected and sent back.

By 7.15 p.m. the whole objective had been captured, and consolidation had started. The line was approximately the house at Q26a61—Q26b71—Q26d89—Q27c19—Q27c36—Q27c10—Q32d77.

Companies were a good deal mixed owing to the part that A Company had taken in the attack, and to the fact that the objective was semi-circular. Most of the support Company had to be used to fill gaps in the line, and the only troops now in reserve consisted of one platoon of A Company. A request was therefore sent for two platoons of the Rifle Brigade to be sent forward. These platoons arrived shortly before dawn on the 15th, and dug in alongside the reserve platoon of A Company on the line of the light railway, just South-West of the village.

During consolidation the enemy heavily shelled the village and its Western approaches. As, however, our line for the most part was clear of the village, little damage was done.

At dawn, on the 15th, a number of Germans in full marching order, estimated at least 200 strong, were seen to be advancing in twos and threes into shell holes from Q20 central, and the houses to the South-East of it; there was also considerable movement about the groups of houses at Q26b95, and at the Northern end of the Bois de Pacaut.

Every available rifle and Lewis gun were brought to bear on all Germans seen, and severe casualties were inflicted on them. If a serious counter-attack was intended, it was nipped in the bud and never developed.

The German stretcher-bearers were busy for the remainder of the day.

During the 15th information was received that the 10th Brigade would attack the Bois de Pacaut from the South at 5.40 p.m., and that the Battalion would co-operate, joining up with the left of the Duke of Wellington's Regt., and finally capturing the Riez-Le Cornet Malo road.

Light Company on the right was holding from the Canal to Riez, facing the Western side of the Bois de Pacaut.

The following orders were issued:—

The right platoon of Light Company, just North of the Canal, was to left form at zero, connecting up with the Dukes on the edge of the wood. As this platoon got level with the left of the next platoon, the latter was also to left form, and the Company would then advance on a two-platoon front, the other two platoons similarly left forming, when the time came, and following on in support.

The Dukes crossed the Canal, however, at Q32a60, and attacked the wood from the West, coming under heavy artillery fire and suffering serious casualties.

It was too late to send further orders to Light Company, and the two right platoons carried out the instructions given them. They moved out at zero, but finding no one on their right as expected, they attacked most gallantly straight towards the wood, and actually entered it alone about Q33a83. They met serious opposition, and came under heavy fire, and being entirely unsupported were unable to stay there: they finally fell back, and dug in with the Dukes on the approximate line Q33a16—Q33c57. These two platoons had about 50% casualties.

Companies and battalions were now a good deal mixed, and the night of the 15th-16th was chiefly devoted to re-organising.

At noon on the 16th the enemy opened a heavy trench mortar and artillery fire on our front line, the village and the Western outskirts. Germans, at least 200 strong, were seen to be massing around Q26b85, Q26b99 and Q27 central.

Excellent work was now done by Captain Boucher: he at once rushed up to the front line, rallied the front posts which were beginning to waver, and pushed forward his supports. The situation for a time was threatening: the left of our line was unable to come into action owing to accurate machine gun fire from the various buildings.

The enemy, carrying full marching order, attempted to debouch between the two buildings at Q26b99 (not shown on the map), past the forked roads at Q26b85, and through the hedge, Q26b83 to Q27a00.

They were, however, engaged with excellent results by four Lewis guns South of the Riez main street and by rifle fire; heavy losses were inflicted on them, and the attack was broken.

The S.O.S. was sent up; our guns were prompt to answer, and caused further losses to the enemy.

The German stretcher-bearers were again seen to be busy in the afternoon collecting the wounded.

At 2 p.m. the enemy started retiring in twos and threes, and were again heavily fired on.

Thus ended the successful operations of April 14th, 15th and 16th, and the Battalion had every reason to be proud of the results achieved. These results were:—

(*a*) The village of Riez du Vinage was captured, and a bridgehead had been established.

(*b*) One battery of British field guns and one battery of 4.5 Howitzers, lost in previous fighting, were re-captured.

(*c*) Sixteen light machine guns, four heavy machine guns, and a recaptured Vickers gun were taken, making a total of 21. These were all collected and sent back. It was impossible to search the village thoroughly before the relief, and it probably contained a good deal more war material.

(*d*) One heavy trench mortar on wheels was captured in the village.

(*e*) One German battalion, or the equivalent in numbers, was put out of action. One hundred and thirty-five prisoners were sent back to the cages, about 60 dead Germans were counted in the village, and at a low estimate at least 200 other casualties were inflicted by rifle and Lewis gun fire.

This attack proved conclusively the value of section rushes, which had been practised so assiduously at Wanquetin.

Furthermore, the power and value of rifle and Lewis gun fire were definitely impressed on everyone. It was generally conceded that never before had such targets been presented or such execution performed by the bullet. The Battalion's casualties for the period were—5 officers and 210 other ranks.

Officer casualties :—

Died of Wounds—

Sec.-Lieut. L. B. Johnson.

Wounded—

Sec.-Lieut. C. S. Lewis.
Sec.-Lieut. A. G. Rawlence.
Sec.-Lieut. J. R. Hill.
Sec.-Lieut. C. S. Dowding.

During the night, April 16th-17th, the Battalion was relieved by the King's Own of the 12th Brigade, and moved back to billets in Cense la Vallée.

Early in the morning of April 18th, information was received that an attack by the enemy on the XIIIth Corps front was expected, and orders to "stand to" at 5 a.m. were issued.

At 3.45 a.m. the Battalion was ordered to move at once to an assembly position in V6b, South-West of Mont Bernenchon, and to counter-attack immediately any enemy who might have succeeded in crossing the Canal.

The enemy had attacked in considerable force, carrying bridges on which to cross the Canal, but he was beaten back with heavy losses.

Consequently the Battalion was not called upon, and moved back again to Cense la Vallée at 10 a.m.

During the afternoon a further march to Busnettes was carried out.

On the evening of April 19th the 11th Brigade relieved the 10th in the line: the Battalion was in Brigade reserve, three companies and Battalion H.Q. being accommodated in hangars and billets at Lannoy, and A Company on the Canal bank.

As the majority of the three companies were crowded in the hangars without any protection, very great care had to be taken to avoid observation by hostile aircraft.

On the 20th A Company rejoined the remainder of the Battalion at Lannoy.

A further attack by the 11th Brigade was ordered for the morning of the 22nd with the object of extending to the South-East the bridgehead already gained by advancing the line to the Riez-la Pannerie road running through the Bois de Pacaut, a defensive flank being formed from La Pannerie to the Canal at Pont D'Hinges. To this end the following moves took place on the night of April 21st: B Company relieved a Company of the 1st Rifle Brigade on the Canal bank from the road at W4d1o to Pont D'Hinges, and C Company relieved a company of the 1st Hampshire Regt. on the Canal bank from Pont D'Hinges to the stream at W3b16. These two companies came under the orders of the Rifle Brigade.

Light Company moved to Le Cauroy and was attached to the 1st Hampshire Regt.

Battalion H.Q. and A Company remained at Lannoy.

At 5.15 a.m., on April 22nd, the attack took place and was successful, but none of the companies of the Battalion were involved. The enemy's artillery, however, retaliated very heavily, especially on the Canal bank, and B Company in particular had a very uncomfortable time.

The Battalion had the following losses during the day:

Officers :—

Killed—

Captain E. W. Marshall, M.C.
Lieut. D. L. Moore.

Other ranks—25.

In Captain Marshall the Battalion lost one of its most efficient officers, and one who had gained the respect and affection of every officer and man.

The Brigade also had a sad loss in the Brigade-Major, Captain Harston, M.C., who was killed when visiting the new line in the Bois de Pacaut. He had succeeded Captain Prideaux, and had proved himself a most able and popular officer. He was not an easy man to replace, but the Brigade was fortunate in obtaining as its new Brigade-Major, Captain Lester, M.C., another staff officer of the very best type. He at once obtained the confidence of the whole Brigade, and established with all ranks the most cordial relations. He held the appointment with conspicuous success until after the Armistice.

On the following night, April 23rd, the Battalion relieved the 1st Hampshire Regt. and a portion of the 1st Rifle Brigade in the front line from the cross roads Q33d58 to La Pannerie and thence to the Pont D'Hinges.

Owing to the recent fighting the relief was somewhat difficult, and carried out under considerable machine gun fire. Captain H. M. Boucher and Sec.-Lieut. P. Buse were killed.

The Battalion was disposed with A and Light Companies holding the front line, the latter having one platoon on the Canal bank from Pont D'Hinges to W3b81, C Company in support, two platoons on the Canal bank from W3b81 to W3a69, and the other two platoons about W8b98, B Company in reserve at Le Cauroy, and Battalion H.Q. at W2b44.

The next day, April 24th, was fairly quiet, except that the Canal bank received a good deal of shelling: from now onwards the Canal became a favourite target for the enemy's guns.

During the night the dispositions were slightly altered: two platoons of C Company moved up into support in the Bois de Pacaut and the other two platoons took their place on the Canal bank: the position at W8b98 vacated by C Company was taken over by two platoons of B.

A re-adjustment of the line took place on the night of the 25th. B Company moved forward into the front line on the left of A, relieving part of the 2nd Duke of Wellington's Regt. of the 10th Brigade: its left then rested just clear of the Western edge of the wood. The two platoons of C Company on the Canal moved forward into the wood, taking up a position on the left of the remainder of the Company.

Light Company, on the right of the front line, was relieved by a portion of a battalion of the 3rd Division, and took over the position on the Canal bank vacated by C.

The work of consolidation proceeded apace: owing to the marshy ground it was impossible to dig deeper than two or three feet, and this entailed the construction of modified breastworks.

Lateral communications within the wood were opened up, and the rides leading towards the enemy were barricaded.

On the night of the 28th-29th April the Battalion was relieved by the 2nd Essex Regt. of the 12th Brigade, and returned to billets in Cense la Vallée.

The Division now settled down to stationary warfare for the summer. Two Brigades held the line, with one in reserve. At the beginning each Brigade had two Battalions in the front line, with one in support.

On May 8th information was received that the Germans were intending to make an attack on a wide front on the night of the 8th-9th. Consequently each front Brigade put all three battalions into the line, distributed in depth, to be ready for any contingency.

Most detailed information was forthcoming regarding this expected German attack: the bombardment was due to start at 10.30 p.m. and to last for three hours, zero being at 1.30 a.m. The attack was to take place under cover of darkness, and the Germans hoped to reach the Canal at dawn.

This attack, however, never materialised, possibly owing to our exact information and the counter-preparations that were made to meet it.

On the night of the 8th-9th heavy bombardments, lasting for fifteen minutes, took place at 10.45 p.m., and again at midnight, but beyond this the enemy made no move.

The expectation of a German attack remained throughout the summer, with the result that the method of defence was not altered, and all three battalions of each front Brigade were in the line.

To start with, each Brigade did eight days in the line with four in reserve; this was later altered to twelve days in the line and six in reserve.

The Battalion always rested in Cense la Vallée. When in the line on the right it held the sub-sector just West of Facaut Wood, and on the left that facing La Pierre au Beure.

On the whole this period was a fairly pleasant one, though somewhat strenuous.

The thought of a German attack in force was always present, and its extreme probability gave rise to some anxiety. Extra vigilance was therefore imperative, and elaborate defence schemes came into being. Above all else, work was the order of the day, and every man had to toil to the full extent of his capabilities.

The Division arrived in this neighbourhood to find a landscape untouched by war and a complete absence of trenches. On its departure in August, it was able to hand over a thoroughly organised trench system of considerable strength. The front line was continuous and well-wired, and communication trenches had been dug. The trenches nearly everywhere were revetted and duck-boarded, and contained shelters for the men. Water was reached three feet below the surface of the ground, and thus the work necessary assumed larger proportions. By the middle of August considerable progress had been made with breastworks of decided strength in preparation for the winter. Numerous concrete pill-boxes had also been constructed. In addition to the first system of defence, which was excellently designed and strong, other systems were constructed stretching back in depth for several miles.

The most noticeable feature throughout was the wire, which everywhere was very formidable.

The amount of work performed in the time was little short of marvellous, and denoted not only an excellent method, but also the willing co-operation of every officer and man.

During this period any training was impossible except on a very small scale. Brigades were never out of the line for long, and big working parties had invariably to be found. Owing to the importance of the Canal line and the expected German attack, the Reserve Brigade had always to supply two companies to garrison the Canal, these companies coming under the tactical command of the left Brigade in the line. This meant that each battalion of the Reserve Brigade had two companies detached on the Canal bank twice out of every three times when in reserve for half the period that the Brigade was out of the line. This interfered sadly with training.

In spite of this much good work was carried out, especially in regard to musketry by means of 30 yards' ranges.

The heavy programme of work in hand was not allowed to interfere with the usual aggressive tactics adopted by the Division, and the enemy was harassed in every possible way. Our artillery was persistently active both by day and night, and full use was made of concentrations of gas discharged from projectors.

The enemy, on his part, on the whole adopted a policy of keeping his guns quiet by day and showing considerable artillery activity by night. The Canal and its crossings in particular were regularly shelled during the hours of darkness. He also made free use of gas shells, mustard gas being the most commonly employed.

Towards the end of July "winkling" came into fashion, and became a popular form of amusement.

The Germans, possibly because they were expecting to continue their advance at any moment, did not trouble to construct anything in the nature of a trench system. Their front was covered by a number of scattered posts, each holding two or three men.

Daylight patrolling had been actively carried out during the summer by the 4th Division, with the result that the men became skilful and gained confidence, and a thorough knowledge of no-man's land and the enemy's dispositions was obtained. A certain amount of standing corn and long grass existed, and the ground was intersected by shallow ditches; this was all in favour of daylight patrols.

"Winkling" consisted of locating a German post, stealthily approaching it under cover of the corn and the ditches, and returning to our lines in broad daylight with the occupants.

The 12th German Reserve Division was holding this portion of the front, and, judging by the way it lent itself to "winkling," it must have been lamentably lacking in moral and discipline. Between the 24th and 30th of July the Battalion captured in this way eight Germans and killed one Sergt.-Major, who showed fight, all belonging to the 51st Regiment.

A night patrol also lay in wait for Germans approaching one of their night posts, and killed four. These results were obtained without the loss of even one man wounded.

"Winkling" was in favour throughout the Division, but the Battalion easily headed the list of captures.

The German Sergt.-Major, referred to above, took a prominent part in a duel remarkable for the bad markmanship of the duellists. One morning a German post, having been located from our lines, Captain Gough, with Sergt. Winter and another man, decided to "winkle" it. Having worked their way up a ditch, they approached from a flank the post which was in a short length of trench. At the near end of this trench a sentry was posted: he behaved in the usual manner, and was captured without difficulty. Captain Gough, followed by Sergt. Winter, proceeded down the trench, and after a short distance came to a low traverse, the other side of which was a shelter. Out of this shelter emerged a stalwart Sergt.-Major, who was made of sterner stuff than the sentry. For a moment he and Captain Gough glared across the narrow traverse, and then both simultaneously emptied their revolvers at each other, but not a single shot took effect! Sergt. Winter was close behind Captain Gough, and the Sergt.-Major was seconded by a third German peering out of the shelter. The chief actors in this duel, with empty revolvers in their hands, each withdrew a few paces for further consideration, but Sergt. Winter now took action, and with a well-thrown bomb killed the Sergt.-Major and wounded the German in the shelter. The latter then quietly joined the original prisoner, and the party increased by two returned as they came to our lines.

On August 5th, when the 11th Brigade was in reserve, the enemy started to retire. He did so very gradually: his intention was to shorten his front and fall back on the Lestrem line.

It was decided not to press the pursuit, or to make attacks, but to keep him in very close touch and harass him as much as possible.

On the 9th the 11th Brigade relieved the 10th on the left of the Divisional front. The Rifle Brigade took over the outposts on the Brigade front: the Battalion became the left support battalion and the Hampshires the right. The support battalions held our original front line.

In case the enemy was preparing to launch a sudden attack, in the hope of catching us unawares owing to the forward move, it was ordered that in such a contingency the outpost battalion, after breaking up the attack as far as possible, should fall back, and that the original first defensive system should become the main line of resistance. While in support under these conditions, it was found that an excellent training ground was available owing to the lack of direct observation by the enemy, and the possibility of moving freely across country without any of the usual restrictions as to crops. The fullest use was made of this, particular attention being paid to patrol work; its value was apparent later in the outpost line, and when following the Germans as they retired owing to the few casualties incurred.

On August 12th the Battalion relieved the Rifle Brigade in the outpost line, just East of Quentin, two companies holding the front line as picquets, with the other two companies in support, each company being on a two-platoon front. Active patrolling was carried out, and the enemy was found to be in strength.

On the night of the 14th-15th the dispositions were altered, and all four companies were in the front line, being distributed in depth. The enemy was still in strength, and the line was not advanced.

The Battalion was relieved on the 15th by the Hampshires, and became right support Battalion in front of Riez. Work on breastworks was still busily continued, and much useful training was carried out.

On the 20th the Battalion took over from the Rifle Brigade in the outpost line, which had been considerably advanced.

On the following day the enemy was not found in force, and the line was pushed forward to an average depth of 500 yards. The Germans had retired to, and were holding in strength the line of the road running South from L'Epinette.

At 7 a.m., on the 22nd, orders were received that the 74th Division on the left were about to make an attack, and that the Battalion would co-operate as far as possible, but would not endeavour to advance if severe opposition was met. An attempt to push forward Lewis guns was made to cover the advance of the 74th Division, but was not successful, owing to the enemy being in strength, and very little time being available for manœuvring or preliminary arrangements.

The assault of the 74th Division came under heavy machine gun fire and artillery barrage, and failed. The right battalion of the 74th Division fell back several hundred yards behind its original front line, but C Company, on the left of the Battalion, at once formed a defensive flank, and joined up with it.

During the night the 4th Division was relieved by the 19th Division, the Battalion handing over to the 24th Welsh Regt. After the relief the Battalion was conveyed by train from Chelsea Bridge, on the Canal, to billets in Cense la Vallée.

CHAPTER VI

THE CROSSING OF THE COJEUL AND SENSÉE RIVERS—ETERPIGNY—THE DROCOURT-QUÉANT LINE, SEPT. 2ND, 1918—THE ÉCAILLON RIVER, OCT. 24TH, 1918—THE FIGHTING ROUND PRESEAU, NOV. 1ST AND 2ND, 1918

ON the next day, August 23rd, the Battalion was moved in busses to Ames, where only a short stay was made.

At 7 p.m., on the 24th, the transport moved by road to St. Pol, and on the 25th the Battalion marched to Lillers, and entrained for Bryas, near St. Pol. After detraining it marched to billets in Ostreville, whence it again took the road at 6 p.m., on the 26th, for St. Lawrence Camp, near Châtean de la Haie, which was reached at 2.30 a.m. on the 27th.

After a few hours' rest, officers proceeded to the neighbourhood of Monchy le Preux to reconnoitre the ground over which the Canadians had just made a successful attack, and where it was expected that the 4th Division would operate. The Division had now been transferred to the Canadian Corps. (Map 6.)

During the day the only orders received were to the effect that the Battalion would move the next morning, relieve a Canadian battalion at night, and be ready to continue the advance the next morning at dawn.

Early the next morning the Battalion was ordered to relieve the 4th Canadian battalion, and to embus at 2 p.m. Details to be left out of action were selected, and all other preparations for battle made.

The transport moved by road to St. Sauveur, and the details proceeded in busses to Arras.

The original plan was that the bus convoys conveying the 11th Brigade should proceed along the Cambrai road after passing through Arras, and that the Brigade should assemble in the area N4, immediately East of the old Wilderness Camp, which was now completely destroyed.

On reaching Arras, however, owing to the Cambrai road being persistently shelled with H.V. guns, the convoys were diverted and the troops debussed at Blangy. From there the Battalion marched across country to the assembly area.

It had been impossible to make any preliminary arrangements regarding the relief, and it was not even definitely known what line had been reached by the Canadians. Accordingly each battalion sent on ahead from the debussing point a small party to get in touch with the Canadians, and fix up, as far as possible, details for the relief.

Captain E. Paul, M.C., D.C.M., the Adjutant of the Battalion, was sent on for this purpose. It was already getting late, and there was little hope of hearing from him before the time arrived to start. An hour's rest was taken at the assembly area, and teas were issued. No further news was received from the advance party or the Canadians, except that it was reported that the 4th Canadian battalion was in the neighbourhood of Boiry Notre Dame.

The Battalion was ordered to march across country in artillery formation as far as the Bois du Vert, keeping Monchy on the left and the Arras-Cambrai road on the right, and be prepared to carry out the relief without further assistance. After the Bois du Vert a railway running past the North end of the wood to the Western outskirts of Boiry Notre Dame was to be followed.

At 7.45 p.m. the Battalion marched off from the assembly area. There was considerable difficulty in keeping direction: the night was very dark, and a good deal of harassing fire from German H.V. guns was met.

The knowledge of the ground gained during the previous winter was of great assistance, but was lessened to a considerable extent by the German advance in the spring, as a result of which the original trench systems were difficult to recognise, and in addition a good deal of the ground to be traversed had been in German possession, even during the previous winter.

After many difficulties the Bois du Vert was reached, and the railway found, which led to Captain Paul and the Canadian guides. The relief was completed shortly before dawn on the 29th. The men were very exhausted after their long and tiring march.

Sec.-Lieut. R. R. Powell was killed and twenty men became casualties on the way up from shell fire. The Battalion was immediately South of Boiry Notre Dame, with two companies in Lady Lane from O5d28 to its junction with Long Lane, and the other two companies in support just East of Boiry Lane. Battalion H.Q. was at O15a45.

Almost as soon as the relief was over, orders were received for the Battalion to push forward patrols as soon as possible, and endeavour to seize the line of the Sensée River from its junction with the Cojeul River as far as the Moulin du Roi.

The intention of this and the subsequent operations around Eterpigny was to obtain a "jumping off" line for the forthcoming attack on the Drocourt-Quéant trench system.

The task given to the Battalion was far from easy. Nothing was known of the ground to be covered or the enemy's dispositions, and there had been no time for a preliminary reconnaissance. The men were tired and had had no rest.

The Battalion was holding Lady Lane, facing approximately N.E., some 2,000 yards from the junction of the Cojeul and Sensée Rivers: to carry out the task allotted the line had to be swung through a complete right angle to face South-East across the enemy's front. No provision furthermore was made in the orders for the protection of the exposed left flank at the junction of the rivers, and no artillery support was possible.

A, the left front Company, was ordered to push forward patrols to establish a defensive flank from the light railway at O6b05 to the North East corner of Galley Wood. It came under considerable fire, but eventually after some casualties approximately gained the appointed line.

The Hampshires, on the left of the Battalion, in front of Boiry Notre Dame, were notified of the plan of operations, and, as requested, moved forward to join up with the left of A Company.

Light, the right front Company, was ordered to seize the Sensée River from its junction with the Cojeul to P7c61.

B Company, in support on the right, was ordered to secure the Sensée River from P7c61 to the Moulin du Roi.

C Company remained in reserve, to be used as required.

The patrols of Light and B Companies soon came under fire, and progress was difficult.

The enemy was not in very great strength, but there were a number of well-sited machine gun and rifle posts, which were difficult to locate.

At one time it appeared probable that the operation would not be successful, but orders that the objective had to be captured at all costs were issued, and, as was always the case, officers and men left nothing undone to achieve what was asked of them.

The fighting that took place was exceedingly interesting, and resolved itself into the capture one by one of a series of strong points. The time and labour that had been devoted to practices of this sort during previous trainings were amply repaid.

As each machine gun or German post was located, it was at once engaged by rifle and Lewis gun fire, under cover of which men then started to work round to a flank. This manœuvre proved invariably successful: as soon as the enemy saw their flank threatened, they at once retired.

In this way the patrols gradually fought their way forward, until after about 3½ hours they reached their objective. The country crossed was exceedingly difficult. The ground near and between the two rivers was very marshy, and the rivers themselves presented distinct obstacles.

Scarcely a bridge existed, and they had to be crossed either by means of fallen trees or by wading. Every man was wet through before he reached the objective.

The Battalion had carried out an excellent days' work. Subsequently a staff officer in a responsible position said, that if it had been realised what difficulties had to be overcome, the Battalion would never have been asked to carry out such an operation.

Light Company had the heaviest casualties, losing 27 men, with a high proportion of senior N.C.O.'s: A and B Companies each had about half this number of casualties.

During the night an alteration in the dispositions took place with a view to attacking and capturing next day the "jumping off" line. The Rifle Brigade, which had been in reserve, came into the line, relieving a portion of the Battalion from the Moulin du Roi to P7c94.

A Company was relieved on the defensive flank near Galley Wood by machine guns belonging to an independent Brigade attached to the Corps.

The Battalion was then disposed with B and C Companies in the front line, and A and Light in support, just North of the Cojeul River.

The Hampshires became support battalion of the Brigade.

The attack in conjunction with the 10th Brigade on the right was timed for 4 p.m. on the 30th.

The objective given to the Brigade was from the hedge junction P8a05—P8a22—the road 100 yards North-East of the Chapel, thence South in front of the Eastern outskirts of Eterpigny.

The boundary between the Battalion and the Rifle Brigade was a line from P7c94 to the cross-roads P14a69.

THE ATTACK ON ETERPIGNY, AUGUST 30TH, 1918.

The prospects for the attack were not very bright. Little or no sleep had been obtained since the night of the 27th-28th, and the majority of the men had been wet through for twenty-four hours. The efficiency of the Battalion had been further weakened by the large number of senior N.C.O.'s who had become casualties on the previous day.

The support companies dribbled across the Cojeul before zero and assembled between the two rivers.

At 3.45 p.m. our guns started a destructive shoot on Eterpigny wood and village.

Shortly before zero the enemy heavily shelled the area between the two rivers, and a considerable number of men were lost before the start.

At 4 p.m. the creeping barrage fell 150 yards East of the Sensée, and the attack began.

On the left there was little protective barrage, and progress was only possible by section rushes. On the right the barrage was good, and the advance swept forward more quickly. The enemy in many cases did not wait to meet the advancing troops, and retired in front of them, pursued by fire. The whole objective was gained without very great difficulty, but on arrival there the companies were very mixed.

The whole night was devoted to reorganising the companies and consolidating the position in depth.

The Battalion captured 30 prisoners and several machine guns.

The casualties in this action were 5 officers and 190 other ranks.

The following officers were *wounded* :—

Captain G. C. V. Greetham, M.C.
Captain G. C. W. Malet.
Lieut. E. Cockburn.
Lieut. R. W. Ellis.
Sec.-Lieut. N. F. Boyes, M.C.

During the attack, and for some hours afterwards, the enemy put down a very heavy barrage on the line of the rivers, employing a large number of gas shells.

The day of the 31st passed without incident, though the enemy continued to shell heavily the Sensée Valley and in rear of our forward positions.

After dark the Battalion was relieved by the 1st King's Own of the 12th Brigade. The relief was much delayed by heavily shelling with high explosive and gas on the Sensée Valley, and was not completed until 5.45 a.m. on September 1st.

The Battalion marched back to the neighbourhood of Monchy and found accommodation in Spade trench.

Its strength was now reduced to some 300 rifles, and the men were very exhausted.

At noon a Conference at Brigade H.Q. was held, and the attack on the Drocourt-Quéant line, arranged for the following day, was discussed.

PLAN OF OPERATIONS. (Map 7.)

The general idea was that the Canadian Corps, with the 4th Division on its left, would capture the Drocourt-Quéant line, which consisted of two trenches of considerable strength, with a support line in rear, and ran on the Division's front approximately North from Dury. In front of this trench system the Germans had two other lines of defence.

To the 12th Brigade was allotted the task of capturing the Red line, just East of the support trench of the Drocourt-Quéant line.

The 11th Brigade was ordered to follow in support of the 12th, pass through it on the Red line one hour after its estimated time of arrival there, and in conjunction with the Canadians on the right move forward, to secure L'Écluse village and the edge of the Sensée marshes, known as the Blue line.

The village of Étaing was not to be assaulted by the 4th Division, the left of the attack passing a little South of the village, which was to be kept under a heavy bombardment and a smoke screen.

Simultaneously with the advance of the 11th Brigade from the Red line, the 12th Brigade was to send parties to work Northwards up the trenches towards Étaing.

Arrangements were made to protect the left of the Division by pushing forward machine guns North of the Sensée, from the direction of Galley Wood.

Tanks were detailed to co-operate in the attack. Zero was timed for 5 a.m.

The 11th Brigade was ordered to advance with three battalions in line, from left to right the Somerset L.I., the Hampshires and the Rifle Brigade, with the 2nd Seaforth Highlanders, lent by the 10th Brigade, in support. The remainder of the 10th Brigade remained in reserve.

The Attack on the Drocourt-Quéant Line, Sept. 2nd, 1918.

At 11.45 p.m., on September 1st, the Battalion marched off to a position of assembly in P13 central, guides having been previously sent on ahead to reconnoitre the route. It arrived in position at 2.30 a.m. on the 2nd, and at once proceeded to dig in.

Almost immediately and before the men had obtained cover, a heavy burst of shelling took place, and some twenty casualties were sustained.

Sec.-Lieut. A. M. Wright was killed, and Lieut. F. J. Clark wounded.

This shelling continued off and on for the remainder of the night, but contrary to expectation slackened off after the attack had actually started.

At zero—5 a.m.—the assault began, and good progress was made.

At 5.30 a.m. the Battalion advanced in artillery formation to the first German trench, where an halt was made for an hour. The advance was then continued to the second German line, where another halt was made under fairly heavy shell fire.

At 7.20 a.m. the advance to the Red line was continued: the correct direction had been kept, but there were no signs of the 12th Brigade in front.

As the Drocourt-Quéant line was approached, the Battalion came under machine gun fire from the left, from the direction of Prospect Farm: the German wire was also very strong, and the advance was temporarily checked.

It was clear that the 12th Brigade had kept too far to the South, and had not cleared Prospect Farm and the ground to the East of it.

By moving further to the right, taking such cover as was afforded by the ground, and advancing by rushes, the leading companies entered the main Drocourt-Quéant line about P15b75 at 8 a.m. This trench was clear of the enemy, but persistent fire was coming from the left flank.

There was still no sign of the 12th Brigade.

The greater part of the Battalion in co-operation with some of the Hampshires on the right now advanced due East with the intention of enveloping the enemy in and to the East of Prospect Farm. In addition to the fire from their left and left rear, they also came under machine gun fire from their right and right front, and it was apparent that the Red line had not been completely captured.

They could not get beyond the Drocourt-Quéant support, with their left about P16a38, where elements of the 12th Brigade were found. Here they stayed for the remainder of the day, pinned to the ground by a close fire.

It was now evident that the left flank was quite in the air, and as there was considerable movement of the enemy among the hedges and banks East of Prospect Farm, a counter-attack did not seem improbable.

Accordingly a mixed party, composed of men of the Battalion and some Hampshires, was sent up the trench to the North to secure the left flank, while two platoons borrowed from the Seaforths advanced over the open in an attempt to clear this locality of the enemy.

Little progress, however, could be made, though a few prisoners were taken, and the situation finally remained with the trench blocked about P9d73.

While trying to put matters right on the left, Captain and Adjutant E. Paul, M.C., D.C.M., was mortally wounded.

By degrees a number of Germans dribbled up until there were some 200 in and about the trench P9d28 to P9b70.

For the remainder of the day the enemy kept up a harassing fire from the left flank, and all movement in the open drew fire.

At 10.30 a.m. a party of the 1st King's Own, of the 12th Brigade, which had previously been detailed to block the main Drocourt-Quéant line, arrived and took over the block established by the Battalion.

The Battalion was reorganised and consolidated the trench from P9d71 to P15b74 with advanced troops in and about the Drocourt-Quéant support line.

The physical condition of the men gave cause for considerable anxiety: they were so exhausted that it was difficult to keep them awake or to get them to carry out the work necessary for consolidating the position.

An exactly similar measure of success had been obtained on the right, and the Canadians had been unable to get beyond Dury.

As soon as it was dusk two companies formed an outpost line by digging posts from the Drocourt-Quéant support line at P16a38 to P9d75.10, while the other two companies held from P15b75 to P9d70.

Practically all the opposition encountered during the day came from the left flank. In addition to having gone too far to the South from its left boundary during the first phase of the battle, as far as could be seen the 12th Brigade made no attack at 8 a.m. Northwards in the direction of Etaing, as had been arranged, with the result that the left flank of the 11th Brigade was continuously exposed.

Some heavy British Tanks were met in the neighbourhood of the Drocourt-Quéant line, and were asked but without success to clear up the situation around Prospect Farm. A few Whippet Tanks would have been invaluable at 8 a.m. for this purpose.

At the time it appeared that the chances of capturing the Blue line would have been greater if the 11th Brigade had been allowed to pass through the 12th immediately the Drocourt-Quéant line had been taken, and if there had been no protective barrage. The Germans were badly shaken, but the pause enabled them to rally and bring up more men.

During the night orders were received that another attack would be made at 5 a.m. on the 3rd, and the advance continued to the Blue line.

All preparations for this were made, but three minutes before zero a runner arrived from Brigade H.Q. with a message saying that the attack had been cancelled. It was just possible to get the necessary instructions to the companies in time, but not before they had deployed and moved forward to get parallel to their objectives.

In view of the physical exhaustion of the men during the previous afternoon and night, the way in which they moved out to the attack was wonderful, and proved yet again the extraordinary spirit and endurance of the British soldier.

At 5.30 a.m., on the 3rd, as the enemy seemed suspiciously quiet, a small patrol was collected and a reconnaissance of the enemy's forward positions was made by the Brigade-Major, Captain Lester, M.C., and the Commanding Officer. No sign of the enemy was found.

Our guns were shelling, but information was at once sent back, and the shelling ceased.

Two strong patrols were immediately pushed forward to secure the ridge on each side of the house at P1cb39, the Brigade-Major accompanying them.

Two companies were sent forward behind the patrols to secure the forward slope of the ridge, and the other two companies followed in support to them.

By 7 a.m. the ridge was definitely secured.

The O.C. 1st Hampshire Regt. was asked and consented to put two companies at the disposal of the Battalion.

These were ordered to pass through the Battalion, one on each side of the house at P10b39 and to advance on L'Écluse, endeavouring to secure the Blue line, and paying special attention to the crossings over the river; they marched off at 7.40 a.m.

A message was sent to the 1st Rifle Brigade asking them to co-operate in the advance on the right in the direction of Récourt Wood; this battalion also moved forward at 7.40 a.m.

Messages were also sent to the 12th Brigade on the left and to the nearest Canadian battalion on the right, informing them of the situation and asking them to co-operate. The Canadians subsequently moved forward, and joined up with the right of the 4th Division.

The Battalion had the satisfaction of being the first to discover the enemy's retirement, and of having advanced some thousand yards before the remainder of the Corps realised the situation.

In the meantime, before the two companies of the Hampshires had arrived on the scene, Captain Lester and Lieut. S. J. Dickinson, with a small patrol of three men, pushed forward towards L'Écluse and captured some prisoners on the slopes leading down to the village. The latter were sent back under escort of the three men, and the two officers went on alone into L'Écluse, where they captured more Germans who had been left behind as a rearguard. These two officers, assisted later by two artillery officers who followed after them into the village, between them cleared L'Écluse of the enemy rear party, captured 25 prisoners, and definitely established the fact that the enemy had retired across the Sensée River.

Their action facilitated a rapid pursuit, and their reports were of great value.

As the advance passed over the ridge some shelling was experienced from batteries on the North of the river. Lower down the slopes machine gun fire was met, also coming from across the river.

The two companies of the Hampshire Regt., on reaching L'Écluse village and wood, found it impossible to establish themselves on the Blue line owing to this machine gun fire. Posts were accordingly placed covering with fire the crossings of the river, while the remainder of the two companies were disposed under cover in the Southern part of L'Écluse village and wood. At this time all movement in the open drew shell fire from the North.

These two companies held the crossings over the river from J35d69 to Q1a24. Touch was established with the Rifle Brigade, which was holding the sunken road in P6d and P12b.

The Battalion, with four machine guns, was holding the forward slopes of the ridge from P4b88—P5a21—P5c43 to P11a18, with Battalion H.Q. in the house at P10b39. Touch was established on the left with the Essex Regt., of the 12th Brigade, in trenches in P4a.

At dusk the posts commanding the crossings of the river moved forward and dug in close to the water's edge. The remainder of the two companies of the Hampshires dug in on the slopes about 300 yards S.W. and S. of L'Écluse village and wood, from P5b98 to P6d19.

During the night the 2nd K.R.R.C., of the 1st Division, took over the crossings of the river, and the troops of the 11th Brigade withdrew to O8, South-East of Monchy.

The relief was completed just before dawn on Sept. 4th. Thus ended a very strenuous period which began on August 28th with the relief of the Canadians at Boiry Notre Dame. Since then the Battalion had been continuously fighting, and had in addition covered considerable distances. Throughout the time only broken snatches of sleep of short duration had been possible.

The casualties during the period amounted to 15 officers (four remaining at duty) and 312 other ranks.

Officer casualties, in addition to the five wounded at Eterpigny, which have been already given :—

Killed—

Sec.-Lieut. R. R. Powell.
Sec.-Lieut. A. M. Wright.

Died of Wounds—

Capt. & Adjt. E. Paul, M.C., D.C.M.

Wounded—

Captain C. W. Hall, M.C.
Lieut. F. J. Clark.
Sec.-Lieut. J. Measures.

After a few hours' rest the Battalion marched off at 1 p.m. to Tilloy les Moiflaines, where it embussed and proceeded to billets in Cambligneul.

A fortnight's rest was now obtained, and the usual training and refitting were carried out.

On Sept. 10th, to the great grief of all, Captain E. Paul, M.C., D.C.M., died of wounds in the hospital at Aubigny. He was buried with military honours the following day. His death was a sad loss to the Battalion. Early in the war as a Regimental-Sergt.-Major, and later as Adjutant, he had earned the respect and affection of every officer and man who had served with him. Time after time he had proved himself a very gallant gentleman and an officer of exceeding ability.

The command of the 11th Brigade now passed to Brigadier-General W. I. Webb-Bowen, D.S.O., who remained in command until demobilisation started after the Armistice.

On Sept. 19th the 4th Division returned to the line and relieved the 11th Division.

The Battalion marched at 6.45 a.m. to Mingoval, and from there proceeded in busses to Feuchy Chapel Cross Roads. A halt for dinners was made West of Monchy, and the Battalion then moved forward into Brigade reserve, relieving the 6th Lincolns in O2c & d, East of Monchy. A quiet time was spent in this position, and it was possible to continue training.

On the 24th the 11th Brigade moved back into Divisional reserve: the Battalion on relief by the 2nd Essex Regt. proceeded to Orange Hill. Very little accommodation existed here, and a good deal of work was necessary to make shelters and bivouacs for the men.

A draft of 136 men arrived: this made an approximate addition of 200 to the strength of the Battalion since the last operations.

Owing to the possibility of a German retirement, the Reserve Brigade was ordered to be ready to move forward at an hour's notice in support of the leading Brigades.

On Sept. 29th the Divisional front was extended to the right along the marshes as far as the Écourt St. Quentin-Palluel road. This was effected by the Battalion moving into the line on the right of the 10th Brigade, and relieving the right front company of the 2nd Seaforths and three companies of the 56th Division.

The Battalion was disposed with A and B Companies in the front line, with posts on the water's edge, its left being just clear of L'Écluse.

Right Company was left support company near the Bois de Récourt, and C right support company just North of the Western end of Écourt St. Quentin. Battalion H.Q. was in the Quarry at Q13c19.

On the following night the remainder of the 11th Brigade came into the line, and relieved the 10th.

On Oct. 2nd the Battalion's frontage was shortened: B Company, on the left, was relieved by a company of the Rifle Brigade, and moved back to the neighbourhood of the Quarry in Q13c19.

A particularly quiet and pleasant tour was spent in this locality. The Sensée marshes had been considerably flooded by the Germans, and the whole front was completely covered by water. The opposing lines were some distance apart, and separated by this water.

A home-made boat was improvised for the purpose of carrying out patrols, but did not prove very seaworthy. With the exception of occasional bursts around Battalion H.Q., there was practically no shelling.

During the seven days spent here, Lieut. H. V. Berry, wounded by an aeroplane bomb, was the only casualty.

On Oct. 6th the Battalion was relieved by the 10th Canadian Battalion, spent the remainder of the night on Orange Hill, and on the following day moved to billets in Berneville, where it was joined by a draft of 92 men.

Much to the regret of all, Major-General Matheson was now transferred to the Guards' Division. Under his able leadership the 4th Division had reached a high state of efficiency, and had achieved a continuous series of successes both in attack and defence. It was fortunate in obtaining as his successor Major-General Lipsett, who had a brilliant record as Commander of a Canadian Division; but this distinguished officer's career was brought to an untimely end very soon afterwards. Only a few days after assuming command of the Division, he was killed by a sniper when visiting with his usual disregard of danger the outposts on the exposed hillside in front of Haspres. Major-General Lucas succeeded him, and remained in command until the Division returned to England.

A short stay only was made here, and on Oct. 11th the Battalion moved by bus to Fontaine Notre Dame, where accommodation was found in old German shelters.

On the 13th it marched to Escaudœuvres, on the outskirts of Cambrai.

On the 17th the 4th Division, now forming part of the XXII. Corps, relieved the 49th, the front line being on the forward slopes leading down to the River Selle and facing Haspres and Saulzoir. (Map 8.)

The Battalion relieved the 1/5th York and Lancaster Regt., the right support battalion, in the area O33c & d. No cover existed, and the men dug in on open ground.

The enemy had been recently counter-attacking with Tanks, and every preparation was made to deal with such an occurrence. Captured German anti-tank rifles were much in request, but owing to their excessive weight it was almost impossible to carry them about.

The principle was now laid down that sections of field artillery should follow the attacking infantry as closely as possible with a view to engaging directly any hostile Tanks that might appear: they were often of use also in knocking out machine guns.

An attack by the 11th Brigade on the village of Haspres was arranged for the morning of Oct. 20th. The Battalion, which was in reserve, was ordered to supply 2 officers and 152 other ranks to accompany the assaulting troops and to carry bridges, machine gun and trench mortar ammunition.

At dusk, on the 19th, the Battalion moved forward in heavy rain to the area O28a, just South of Avesnes-le-Sec, and dug in.

At 11 p.m. the attack was cancelled, owing to the enemy having withdrawn his advanced posts.

Oct. 20th was again very wet. At 9.48 a.m. the Battalion was ordered to capture the village of Haspres from the South-East. The 51st Division on the left was reported to have reached O6c. On the right the 10th Brigade had captured Saulzoir, and had established an approximate line P20 central—P14d30—P15a00—P15d00—P22a27—P22 central—P23b01. The Battalion had to advance along the valley in O29 central, South of the woods in P25, through Saulzoir to positions of assembly in the sunken road in P21d and the sunken road in P22c and P28a, and then attack through the 10th Brigade.

Zero was ordered for 12.30 p.m.: 6,000 yards had to be covered to reach the assembly positions, so very little time was available.

From the Mill P20c26 to P15a00 was selected as the jumping-off line for the attack, and it was arranged that the creeping barrage would fall 200 yards in front of this line at zero. B and C were detailed as the leading companies, B on the right; with A and Light Companies in support, A on the right.

The railway was taken as the dividing line between the companies. B and C Companies were ordered to go straight through to the objective, and Light Company was detailed for mopping up the village. A Company was ordered to right form in succession, as the advance proceeded, and to take up a position from the road at P8a23, joining up with the left of the 10th Brigade at P15doo.

As the Battalion passed through Saulzoir, where a number of French civilians were seen, it came under heavy shell fire, but luckily had no casualties. It only arrived in position at 12.25 p.m., five minutes before zero, and moved straight to the jumping-off line.

The position in the village of Haspres was somewhat uncertain, and it was not known if the Germans were holding it in strength. If the Hampshires in the meantime had been able to send patrols across the River Selle into Haspres, it was understood that the attack would be cancelled, though it would be impossible to notify the battalion in time. No definite orders had been issued on the subject, but in order to be prepared for such a contingency the companies were ordered in the event of no barrage falling at zero to push forward at once strong patrols into the village without waiting for orders, and to endeavour to secure the objectives.

It so happened that the attack was cancelled, as the battalion on the left reported that its patrols were in the village. This information later proved to be incorrect, and Brigade H.Q. ordered the attack as originally planned to take place at 3.30 p.m. By the time these orders were received the Battalion was well inside the village, and was in danger of being caught in our barrage.

A mounted orderly galloped back to Brigade H.Q. and fortunately arrived in time to enable the barrage to be cancelled at the last moment.

The Battalion pushed forward rapidly, and without any serious opposition secured the original objectives and established the line P15c19—P14a95—P8ac2—O12d68—in touch on the right with the 10th Brigade.

A and B Companies held the front line, with C and Light in support: Battalion H.Q. was established at P19b19.

A considerable number of the enemy were seen retiring over the hill, and were engaged with fire.

During the night constant patrols were sent out to locate the enemy, and at first he was found in strength. At 4 a.m. a German post was surprised about P8a28: a machine gun was captured, and its crew of four men killed. Soon afterwards it was discovered that the enemy was continuing his retirement.

At dawn on Oct. 21st the Battalion advanced in close pursuit of the enemy, and dug in on the line P2b87—P3c50—P9 central—P9d73—in touch with troops on both flanks. Patrols were sent forward, but were unable to make ground owing to rifle and machine gun fire. In front of the line now held the ground sloped steeply down to the River Ecaillon, and was very exposed and devoid of cover. The enemy were in strength on the opposite slopes, and had advanced posts West of the river. During the afternoon the enemy heavily shelled the top of the ridge in rear of our positions.

At dusk strong patrols were sent out to occupy the sunken road running from the cross roads at J33b33 through P4a & c.

The right patrol found the Ferme de Bouveneule strongly held, but at the third attempt drove the enemy out, losing four men in so doing.

After this farm had been captured, the enemy heavily shelled it for an hour.

The left patrol in the meantime established itself without much difficulty in the neighbourhood of the cross roads at J33b33.

The front companies then moved forward, and took up a position along the sunken road. One platoon on the right formed a defensive flank from P10a43 to P9d73 in order to maintain touch with the 10th Brigade, which had not moved forward.

The support companies were established on the top of the ridge, with Battalion H.Q. at P7d53.

The Battalion was relieved during the night by the 1st Rifle Brigade, and withdrew into Brigade reserve in Haspres. The casualties during these operations were not heavy, and totalled 27.

After a short period of rest the Battalion marched forward out of Haspres at 11.45 p.m. on Oct. 23rd to take part in an attack to force a crossing of the River Écaillon at 4 a.m. on Oct. 24th.

PLAN OF OPERATIONS.

The 4th Division was attacking in co-operation with the 51st Division on the left and the 19th Division on the right. The 11th Brigade was attacking on the left and the 10th Brigade on the right of the Divisional front, with the 12th Brigade in reserve.

The Battalion was the right and the Hampshires the left battalion of the 11th Brigade, with the Rifle Brigade in reserve.

The Brigade was to assault from the sunken road West of the river, which had been secured by the Battalion on the evening of the 21st.

The junction between the Hampshires and the Battalion was P4c12.

The left boundary of the Brigade ran from the Northern outskirts of Monchaux to the cross roads at K25a23, and the right boundary along the La Layette-Le Taravise Mill road, inclusive, as far as P5b50, thence to K31c45.65.

The 11th Brigade was given three objectives :—

The first, or Blue line, approximately along the line of the road from Monchaux through P5 central.

The second, or Yellow line—J29d24—J36c33—Q1a50.

The third, or Green line, approximately the road from the cross roads at J30d78 through K31a & c.

It will be seen that the frontage of the Brigade gradually decreased as the advance proceeded.

Accordingly the Hampshires were given the task of capturing Monchaux, and were not asked to attack further.

To the Battalion was allotted the task of capturing the Yellow and Green lines on the whole of the Brigade front.

The details of the plan of attack were as follows :—

The leading companies, C on the left, Light on the right, were ordered to capture the first objective, or Blue line.

The support companies, A on the left, B on the right, were ordered to leap-frog through the leading companies at the first objective and push right through to the final objective, or Green line.

Definite parties were detailed to mop up the Ferme de Pluvinage, La Layette, Le Taravise Mill, and neighbouring houses, and also the Hermann Stellung, a German defensive line of unknown strength, just in front of the first objective.

Two platoons of the Rifle Brigade were attached to the Battalion : they were ordered to advance immediately in rear of A and B Companies, and were given the task of mopping up the Yellow line. Having done this, they were instructed to consolidate the Yellow line astride the road junction in J36c. After the capture of the Yellow line, C and Light Companies were ordered to move forward from the Blue line and consolidate the Yellow line, one Company on either side of the two platoons of the Rifle Brigade.

The Hampshire Regt. was ordered to consolidate with three companies across the whole Brigade area in J35a, J35c and P5b. The fourth company was ordered to consolidate on the left of C Company in J36a, with one platoon thrown forward to the cross roads at K25a23 to join up with the 51st Division.

The Rifle Brigade had orders to advance and consolidate the first objective as soon as the Yellow line had been captured.

A and B Companies were told to send forward patrols after the capture of the Green line to exploit success and join up with the 10th Brigade, which had as its furthest objective the village of Querénaing.

Four bridges were to be carried forward on which to cross the Écaillon river, two with the leading platoons of C Company and two with the leading platoons of Light Company. Carrying parties for these bridges were supplied by the Rifle Brigade, and came under the direct command of the O.C.'s Light and C Companies.

In order to closely support the advance, and to guard against a hostile counter-attack with Tanks, a section of field artillery was ordered to move forward as soon as possible after the Yellow line had been captured to a position about J36c50.

The method of assembly was as follows :—

C Company assembled in the sunken road with two platoons North and two platoons South of the Ferme de Bouveneuie. Two platoons of Light Company occupied the sunken road from the right of C Company as far as the bend in the road at P10c59, while the other two platoons assembled on a tape in prolongation of the road and joined up with the 10th Brigade about P10c75.

A Company assembled on a tape from P3d40 to P9b54, and B Company on a tape from P9d80 to P10c03, the bottom of the valley being avoided, as it was likely to be shelled.

The creeping barrage was arranged to fall 200 yards in front of the assembly positions of C and Light Companies, advancing at the rate of 100 yards in four minutes till it reached a line 200 yards beyond the first objective, where it dwelt for 50 minutes, and subsequently moving forward at the same rate to the final objective, with a halt of 10 minutes in front of the Yellow line.

The Battalion left Haspres at 11.45 p.m. on Oct. 23rd, and moved forward into the assembly positions. The night was quiet, no difficulty was encountered, and the troops assembling on tapes at once dug in.

The Rifle Brigade, which was holding this part of the front, was not relieved in the ordinary way. As soon as the Battalion was in position the Rifle Brigade withdrew, and their advanced posts in front of the assembly positions were not relieved. Each of the leading companies, however, pushed forward posts for their local protection, but withdrew them before zero.

Battalion H.Q. and the Regimental Aid Post were established under the bank at P9c30. The Battalion was in position by 2 a.m. on Oct. 24th.

The Attack on the Écaillon River, Oct. 24th, 1918.

At 4 a.m. the attack started behind an excellent barrage, and no serious opposition was met until the river was reached. This proved to be a considerable obstacle, 7 or 8 feet wide, and about 5 feet deep, and had wire on both banks and some in the stream itself.

The left bridge was thrown across without much difficulty, but the other three gave a good deal of trouble, and a distinct delay occurred. The Germans now opened fire from the slopes across the river. The bridges were eventually got into position by men wading out into the middle of the stream. While this was taking place, the German fire was neutralised to some extent by good covering fire from our bank of the Écaillon.

After crossing the river the advance continued rapidly to the Blue line: La Layette and the houses round Le Taravise Mill did not cause much difficulty. The Hermann Stellung line, however, in front of the first objective was found to be well wired, and consisted of a series of isolated posts and short lengths of trench. It was strongly held by machine guns and some trench mortars. The leading companies, by the free use of rifles and Lewis guns, and by working round the flanks, quickly overcame the enemy resistance, and the Germans, who at first showed fight, suffered heavy casualties: the survivors soon surrendered.

After this line had been captured, the leading companies swept on and assaulted the Blue line with the bayonet.

The support companies now passed through the leading companies towards the Yellow line. Various machine gun posts were met and dealt with as before most successfully by fire and outflanking movements. A large number of the enemy were surprised and captured in the sunken road about P5b83.

After the Yellow line had been captured it was at once consolidated by C and Light Companies in touch with the Hampshires on the left and the 10th Brigade on the right.

The further advance of A and B Companies to the Green line met with serious difficulty. The 10th Brigade had been unable to get beyond the Yellow line, and these companies came under heavy machine gun and rifle fire from the right. Movement in the open became almost impossible, but in spite of this A and B Companies persisted in their efforts to reach their objective, and eventually a mixed body of about 80 men of the two companies, under Lieut. J. A. Radford, reached the Green line. Here, quite unexpectedly, the Germans were found to have a line strongly wired. Further progress was impossible, and this gallant party dug in 40 yards from the German wire, approximately from J30d81 to K31a05.50. They held on here for the remainder of the day, though their right flank was completely in the air. On the left they established touch with the weak elements of the 51st Division, which had also reached this line.

The left of A and B Companies rested on a communication trench, running from the Green line towards the Yellow line. As soon as it was dusk the enemy made a strong bombing attack down this trench, but it was beaten off.

During the day German low-flying aeroplanes showed great temerity, and in addition to making all movement in the open dangerous, inflicted a number of casualties on our men by machine gun fire.

At dusk a company of the Rifle Brigade was sent forward to form a defensive flank by establishing posts from the right of A and B Companies on the Green line to connect with the 10th Brigade.

Before the attack the Battalion had received a promise that it would be relieved after dark on Oct. 24th. Before this took place, however, orders were received for the advanced troops to fall back and take up a position 300 yards from and parallel to the Green line, preparatory to a further attack by the 12th Brigade on the following day.

In view of all the circumstances this was no easy matter, but it was successfully carried out, and the line as ordered was handed over to the incoming troops.

During the night the Battalion was relieved by the 2nd Essex Regt., and returned to Haspres. Thus ended a successful day, carried out under very propitious circumstances. Previous to the attack the men had a good rest in comfortable quarters, and before marching off to the line were given a hot meal. Only two hours were spent in the assembly area, and they had been promised a relief at the end of the day. Their spirits and moral were therefore excellent, as was reflected in the dash shown in the attack. It is impossible to state the number of prisoners taken in this action by the Battalion, as no definite record was kept, and many were sent back through the 10th Brigade: at least 150 must have been taken, and probably more.

Similarly a large number of machine guns and trench mortars of various kinds were captured, but it was not possible to count them accurately.

A great many Germans were killed, and the number of dead to be seen was far above the average.

The Battalion lost 8 officers, in addition to four who were wounded but remained at duty, and 149 other ranks.

Officer casualties :—

Died of Wounds—

Sec.-Lieut. P. E. Austin.
Sec.-Lieut. A. J. Gardiner.
Sec.-Lieut. L. A. Young.

Wounded—
Lieut. E. Cockburn, M.C.
Sec.-Lieut. R. P. C. Smith.
Sec.-Lieut. M. Hetherington.
Sec.-Lieut. H. T. Trew.
Sec.-Lieut. T. C. Hackwell.

On Oct. 28th the Battalion once more moved forward, relieving the 2nd Essex Regt. in Brigade support. Light and B Companies held the railway embankment in K27d, A and C were disposed in K31b & d, and Battalion H.Q. was in Delferrière Farm at K26b22.

Since Oct. 24th Artres had been captured, and the line ran along the River Rhonelle.

A further attack had been planned for Oct. 30th, but it was eventually postponed until the morning of Nov. 1st.

In the evening of Oct. 31st A and C Companies moved up to the railway embankment, which had been heavily shelled and gassed during the day.

Plan of Operations.

The general plan of the attack was that the 11th Brigade should attack on the 4th Division's front to capture Preseau and the high ground to the North of the village in conjunction with the 61st Division on the South and the 49th Division on the North.

The objective given to the Brigade was L14c91—L14a45—L8c00—L7 central—L7a50—K6d55. Its boundary on the right was K29a47—L19b15—L20a07—L14c91, and on the left the halt at K21c00—K15c00—K12a00—K6d55.

The following was the method of the attack:—

The Hampshire Regt., plus Light and A Companies of the Battalion, attacked on the left, and the Rifle Brigade, plus B Company of the Battalion, on the right.

The Battalion, now reduced to Battalion H.Q. and C Company, with two companies of the 1st King's Own of the 12th Brigade attached, was in support, and assembled along the railway embankment from the Halt at K21c00 to K27d55.

Light, A and B Companies assembled with the battalions to which they were attached, and came under their orders from the time of assembly.

Four footbridges had already been thrown over the River Rhonelle at Artres, and four more were allotted to the Hampshires. A Company was given the task of carrying forward these bridges, and placing them in position for the leading companies.

Light Company was ordered to follow in close support of the Hampshires as far as the road running through K12c & d, where they were to leap-frog through and capture the final objective.

B Company was ordered to support the Rifle Brigade, and push home the attack if required. After the village had been captured it was to consolidate on a line from the Old Mill to the Church on the South-Western outskirts.

The Battalion was ordered to move forward as soon as the objective had been captured and consolidate the line K17c - K23a, establishing touch on both flanks with the support battalions of the 49th and 61st Division.

THE ATTACK ON PRESEAU, NOVEMBER 1ST AND 2ND, 1918.

The attack started at 5.15 a.m. on November 1st, behind an excellent barrage, and the objective was reported captured at 7.45 a.m.

On the right little opposition was met at the start, but about 1,000 yards from Preseau a strong point was encountered, which for a time delayed the advance.

B Company pushed forward into the front line, and the resistance was overcome by rifle fire, but not before Captain L. A. Osborne, D.S.O., M.C., the Company Commander, had been severely wounded.

The advance then swept on, and B Company was the first to enter the village, where the Germans were holding the houses in strength. B Company dug in on the line from the old mill to the Church, as ordered, while the

Rifle Brigade went through the village to the final objective.

At 9.30 a.m. the enemy put down a heavy barrage, and counter-attacked our right flank. The 61st Division was forced back, and eventually the Rifle Brigade had to give up the village, and, retiring through B Company, took up a position 300 yards West of Old Mill Spur. B Company later conformed and moved back into position on the right of the Rifle Brigade: touch with the 61st Division was not secured till late in the day.

On the left the advance proceeded successfully, and Light Company leap-frogged through the Hampshires as arranged. A line was consolidated along the road just short of the final objective from L7a00—L7d00.

At 4 p.m. the enemy carried out another counter-attack in force, this time on the left.

The 49th Division gave ground, and the posts of Light Company were driven back 200 yards.

Captain P. G. K. Harris, M.C., assisted by Coy.-Sergt.-Major R. Johnson, then rallied the men, and ordered them to charge. The whole line sprang forward with a cheer, flung the Germans back, and re-occupied the original position.

A few prisoners and a machine gun were captured, and a considerable number of Germans killed. Our barrage which fell in response to the S.O.S. accounted for a good number more.

The line subsequently was withdrawn to conform with the situation on the right.

In the meantime the Battalion had moved forward and taken up its allotted position in K17c and K23a.

After dark the 1st King's Own relieved the Hampshires and the 2nd Seaforth Highlanders relieved the Rifle Brigade. On completion of the relief the Hampshires and the Rifle Brigade had orders to move back into Artres, and the Battalion to concentrate in K17.

The King's Own and the Seaforths came under the orders of the 11th Brigade.

At 10 p.m. orders were received that the Battalion would be in support to the Seaforths who, in conjunction with the King's Own, were to carry out another attack on Preseau at 5.30 a.m. on the next morning, Nov. 2nd. Instructions were at once sent to the four companies to move immediately to a position in K24a & b, 300 to 400 yards in rear of the Seaforths, who were assembling on the line K18d35.95—K18d54, thence due South to K24d59.

When these orders were sent the exact positions of Light, A and B Companies were not known, as they were in course of being relieved in the front line. As a result, C Company alone received its orders in time to dig in on the assembly area before zero. Light and A Companies only reached their assembly positions as the attack started.

B Company, which was not relieved till very late, did not receive its orders until after zero.

The objective and the brigade boundaries for Nov. 2nd were exactly the same as on the previous day.

The Battalion's rôle was to move close behind the Seaforths and mop up Preseau. The failure to thoroughly clear the village in the first attack was largely responsible for the inability to hold it. In addition to closely supporting the Seaforths, the Battalion was ordered to drive home the attack if it was held up, and after clearing Preseau to consolidate on the Eastern outskirts of the village.

At zero the attack started behind a heavy 18 pdr. barrage, and the Battalion, less B Company, moved forward into Preseau in rear of the Seaforths. Although the Seaforths had passed through to the objective, many snipers were still active in the village, and nearly all the cellars contained large numbers of Germans, who in most cases readily surrendered and were sent back. The snipers, however, hidden in the houses caused a considerable number of casualties, and were difficult to deal with.

The task of mopping up was a big one, and the village was not completely cleared for some hours.

At 9 a.m. B Company joined the other companies, and finally the whole Battalion dug in on the East of the village.

At 9.45 a.m. it was ordered to withdraw to a line K18d30.95—K18a55, thence due South to the Divisional boundary. The Battalion had consolidated this line by 1 p.m. in touch with the Rifle Brigade on the left and a battalion of the Oxford and Buckinghamshire L.I. on the right.

After dark it was relieved by a company of the 1/9th West Yorkshire Regt. of the 11th Division, and moved back to Haspres.

During the operations on Nov. 1st and 2nd the Battalion lost 5 officers and 88 other ranks.

Officer casualties.—

Killed—

Captain C. C. H. Lewin, M.C.
Sec.-Lieut. A. P. Bower.

Wounded—

Captain L. A. Osborne, D.S.O., M.C.
Sec.-Lieut. F. W. Perrett.
Sec.-Lieut. C. G. Harrison.

During the clearing of Preseau, Captain P. G. K. Harris M.C., was the chief performer in an incident which gave rise to some merriment. He was standing at the top of some cellar steps collecting prisoners, when a German came up from below "kamerading" with such enthusiasm that he collided with Captain Harris and knocked him down. Captain Harris sat down violently on top of a dead German, and in his efforts to rise put his hand on the dead man's face. This was too much for Light Company's Commander; he leapt at the offender and, mindful of his Oxford days, caught him such a left under the jaw that the unhappy German did not recover consciousness for a long time.

In the two days' fighting the 11th Brigade, with the two attached battalions, captured 43 officers, 1,655 other ranks, 5 field guns, and 2 Tanks, besides machine guns and trench mortars too numerous to count.

After the fighting round Preseau, Lieut.-Colonel Earle, D.S.O., commanding the 1st Hampshire Regt., sent the following message to the Battalion:—

"I cannot adequately express my appreciation of the manner in which Light Company and A Company supported me and carried out my orders during the operation. I can never hope to have the honour of commanding better men."

This was the last fight in which the Battalion took part, as it remained in Haspres until the Armistice on Nov. 11th.

The 4th Division did not form part of the Army of Occupation in Germany, but remained in France and Belgium. After the Armistice the Battalion was billeted close to the scene of its last fight in the village of Curgies, a few miles from Valenciennes.

On Jan. 5th, 1919, it moved to Haine-St.-Pierre, in Belgium, 15 miles East of Mons, and later to Binche.

The Battalion returned to England in the middle of June, 1919, as a cadre, consisting of Captain & Adjutant E. H. C. Frith, Captain & Quartermaster A. Neale, O.B.E., Lieut. G. T. Ridge, M.C., Lieut. R. T. E. Green, and some 20 other ranks, in addition to the Band, which had arrived from England two months before. Of these, Captain Neale had the remarkable record of having served continuously since the first day of the campaign without having missed a day's duty, except when away on leave.

For more than four long years the Battalion had been engaged in bitter fighting in France and Belgium; hundreds of brave men had given their lives for their country, many more had received grievous wounds, innumerable acts of daring and gallantry had been performed, and hardships of the most extreme nature had been cheerfully borne.

But this period remains in the memory as a very happy one. A very real feeling of comradeship and true friendship between all ranks was always apparent, and the glorious moments of success and the knowledge of duty well and truly performed outweighed the unhappy days of depression and failure, and atoned as far as was possible for the never-ending loss of well-tried friends. The successes achieved were many and the failures very few.

During the period with which this narrative deals, there were only two occasions on which the Battalion, when attacking as a whole, failed to reach the objective assigned to it and to remain there. These two occasions were July 1st, 1916, in front of Beaumont Hamel, and May 3rd, 1917, in front of Roeux. On the former day the British attack failed for a considerable distance on both sides of the 4th Division, and on the latter the whole attack, extending over a very broad front, broke down and failed to make any progress.

The chief days of success were April 9th, 1917, when the long advance in front of Arras was made; Oct. 4th, 1917, in the Belgian mud by Poelcappelle, when the Battalion alone in the Division reached and held on to its objective; April 14th, 1918, at Riez du Vinage, a purely individual effort carried out by the Battalion alone under the eye of the Divisional Commander, General Matheson, and possibly the most gratifying of all; August 29th, 1918, the crossing of the Cojeul and Sensée Rivers, another individual feat; August 30th, 1918, Eterpigny; Sept. 2nd, 1918, the Drocourt-Quéant line, and the following day, when the Battalion gained a thousand yards' start on the remainder of the Canadian Corps in the pursuit of the retiring Germans; October 24th, 1918, the crossing of the Ecaillon River, when the Battalion alone in the Division reached and remained on its objective.

During all these months every battalion in the 11th Brigade worked loyally together for the general good of the Brigade. Each battalion learnt to esteem and trust implicitly the others, and it would be impossible to wish for more staunch comrades-in-arms than the 1st Hampshire Regiment, the 1st Rifle Brigade, and the 1st East

Lancashire Regiment. It was never the Battalion's lot to be in the front line during any of the main German offensives, but even so it can lay claim to the proud record of never having been forced by enemy action to relinquish any trench or piece of ground on which it had established itself, though on a few occasions voluntary retirements were made in accordance with orders received to conform to the situation elsewhere. Imbued with the Regimental spirit all ranks invariably showed the greatest fighting qualities and maintained a very high state of discipline which alone made it possible to achieve such results. This discipline was not the outcome of the moment, but the traditional discipline of the Regiment, founded on the mutual respect which all ranks had for one another, and even during the months immediately after the Armistice, which possibly provided the most searching test of the whole war, it never relaxed in the slightest degree. Tradition is a wonderful thing, and throughout it shone brightly as a guiding light to those to whom the honour of the Regiment was for the time being entrusted. In the darkest days, and after the most crushing losses, there were always a few imbued with the Regimental spirit, and all for which it stands, to carry on and to impart their knowledge to the more recent arrivals.

One and all fully realised that they were building on the foundation so firmly laid by those who had preceded them, and that they were privileged to complete the work begun by former generations in the Regiment. More than any other this war was won by the junior officer and the extraordinary qualities of the rank and file. So long as the same pluck, determination, cheerfulness and self-sacrifice exist, there can be no fear for the future.

APPENDIX I.

Alphabetical List of Officers of the 1st Somerset Light Infantry who were killed or died of wounds between July 1st, 1916, and November 11th, 1918, with the dates of their deaths :—

Sec.-Lieut. P. E. Austin	24/10/18
Lieut. V. A. Braithwaite, M.C.	1/7/16
Captain H. M. Boucher, M.C.	24/4/18
Sec.-Lieut. D. I. Backlake	19/10/16
Sec.-Lieut. N. F. Button	4/11/17
Sec.-Lieut. P. Buse	23/4/18
Sec.-Lieut. A. R. Barnes	4/10/17
Sec.-Lieut. A. P. Bower	1/11/18
Sec.-Lieut. W. Crabtree	21/8/18
Captain C. C. Codner	3/5/17
Sec.-Lieut. H. L. Colville	6/7/16
Sec.-Lieut. S. H. Card	10/4/17
Lieut. T. M. Dodington	1/7/16
Lieut. C. J. O. Daubeny	10/6/17
Sec.-Lieut. R. E. Dunn	1/7/16
Sec.-Lieut. V. F. de Ritter	9/8/16
Sec.-Lieut. J. T. Davies	14/4/18
Sec.-Lieut. C. A. B. Elliott	12/4/17
Captain C. C. Ford	1/7/16
Sec.-Lieut. G. R. Foley	17/5/17
Sec.-Lieut. G. P. C. Fair	1/7/16
Sec.-Lieut. E. R. Foy	11/4/17
Sec.-Lieut. H. J. Griffiths	9/8/16
Sec.-Lieut. E. M. Gardiner	10/4/17
Sec.-Lieut. A. J. Gardiner	24/10/18
Sec.-Lieut. J. A. Hellard	1/7/16
Sec.-Lieut. N. F. Herepath	11/4/17
Sec.-Lieut. A. M. Hill	9/4/17
Sec.-Lieut. J. A. Johnston	1/7/16
Sec.-Lieut. L. B. Johnson	15/4/18

MAP 1.

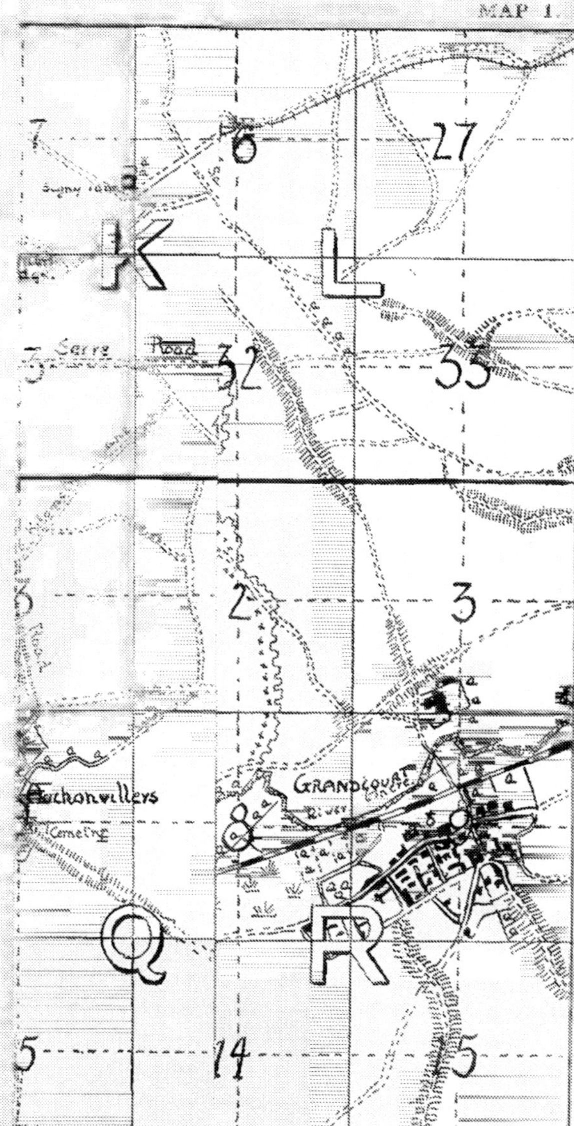

7
6
27
K
L
Serre Road
32
33
3
2
3
Auchonvillers
GRANDCOURT
8
Q
R
5
14
5

Breinigsville, PA USA
10 December 2009
228998BV00001B/181/A

9 780548 784341